Natural Toxins

The Good
The Bad
and
The Deadly

What are they and where do they come from? An examination of the role of toxins in the natural environment and their human uses, both good and bad.

David Wright

Print ISBN 978-1-0685688-3-1

Published by

Llyfrau Cambria Books, Wales, United Kingdom.

Cambria Books and Cambria Stories are imprints of

Cambria Publishing Ltd.

Discover our other books at: www.cambriabooks.co.uk

Acknowledgements.

I greatly appreciate the help and friendship of my colleagues, Professors Pamela Welbourn and Peter Campbell over many years of research and teaching covering many subjects contained herein. I am enormously grateful for their reviews and discussions of this work and suggestions for potentially useful material. I am much indebted to Sarah Hannis for her invaluable assistance with several of the illustrations including the cover design and her help with navigating the complexities of chemical structures.

Above all I thank my wife Susan for her constant support and inspiration throughout.

CONTENTS

Introduction - A Toxic Legacy.

Toxins are generally defined as chemical substances that have an adverse effect on biological organisms. Toxicology is the science devoted to the understanding of how organisms, including humans, become exposed to these chemicals and the nature of the toxic response. Although responses to toxins may be negative and may be hazardous to the health and survival of exposed organisms, toxicology also provides insights into the positive effects of such chemicals and their potential for medicinal and therapeutic use. Many natural toxins have been shown to have beneficial in small amounts, yet in many cases it is clear that the difference between favourable and adverse effects of natural toxins is simply a matter of dose.

What Is a Natural Toxin?

'Natural Toxins' may be simply described as toxins that occur in the natural environment. However, some definitions restrict the term to compounds that are actually derived from living plants and animals as opposed to the physical environment, i.e. soil and water. These chemical compounds have a rich cultural history resulting from their adaptation for human use, although an important part of this narrative relates to the roles that these chemicals play in their natural environment; how they protect and enhance the survival of the organisms that produce them. Indeed, it is the biological properties of natural toxins that often provide the incentive for their extraction and refinement for human use, good and bad.

To illustrate of the difficulty involved in the strict definition of 'Natural Toxins' it is interesting to consider what many consider the 'top three' poisons responsible for human death and sickness, namely Atropine, Cyanide and Arsenic. All originate from widely divergent sources in the natural environment.

Atropine (Figure 1.) is an organic compound found exclusively in biota and is mainly ingested through the consumption of vegetation. It belongs to a major group of biologically active nitrogen-based compounds known as alkaloids. These primarily plant-based compounds have a wide range of biological effects and toxic characteristics, including a long history of fatal human poisonings following loss of blood pressure, respiratory failure, paralysis and coma. However, they are by no means the only plant-based chemical group with poisonous properties.

Cyanides are highly toxic compounds that may also be ingested through the consumption of plant material. However, only a small proportion of cyanide uptake comes directly from biota. For example, a substantial proportion of cyanide ingestion results from the inhalation of atmospheric combustion products of nitrogenous compounds from both vegetation and industrial sources.

Arsenic. It is difficult to talk about toxins in human history without mentioning the metalloid element arsenic which is commonly derived from the earth's crust, including volcano lava and associated gases. Like many other metals such as lead and mercury, arsenic is usually extracted from minerals rather than biological sources, but whose uptake by humans and other animals often results from human activities such as mining, smelting and pollution by industrial waste. Arsenic has also been widely used in a broad spectrum of pesticides although this has been curtailed in recent years following increasing concerns over its toxicity to humans and domestic animals. Human exposure might also include nefarious activities such as murder! Mineral forms of arsenic have been known for over 2,500 years and have been credited with a variety of therapeutic effects including a treatment for syphilis

2

and other diseases. However, it is probably best known as a poison with a unique place in human folklore as well as criminal history!

Following the lead of some early texts on this subject, the history described here includes natural toxins that are of both geological and biological origin and have played important parts in our toxic legacy.

Atropine. An Arbiter of Fate.

Atropine (Figure 1.) is a one of a family of tropane alkaloids that are characterised by nitrogenous bicyclic chemical structures. Tropane alkaloids are found in the plant families Erythroxylaceae, which includes the Coca plant, the primary source of cocaine, and the Solanaceae, a diverse plant family comprising popular vegetables as well as some highly poisonous plant species. Atropine is a crystalline compound that takes its name from the deadly nightshade plant *Atropa belladonna* (Solanaceae) from which it was first isolated by the German pharmacist Heinrich Mein in 1831, although the *de novo* synthesis of the compound did not occur until 1901.

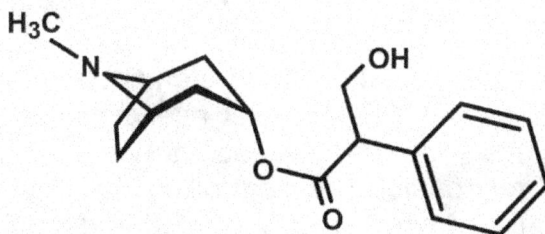

FIGURE 1. Atropine.

However, the toxicity of the nightshade plant's shiny black berries has been known for several centuries, and the generic name *Atropa* is derived from the mythological Greek fate Atropos who holds the shears to cut short human life. The Deadly Nightshade plant is one of several sources of natural toxins found in the Solanaceae, a large and diverse family whose roots can (literally!) be traced back to the early Eocene period in the Patagonian region of Argentina 50 million years ago but is now broadly distributed worldwide. Other toxic members of this plant family include mandrake and henbane.

Many members of the Solanaceae are widely collected and cultivated as food plants, including peppers, aubergines, potatoes and tomatoes. However, the inclusion of the three potentially highly toxic plants deadly nightshade, mandrake and henbane in the same plant family provides a powerful incentive for paying close attention to which part(s) of the plant we choose to consume!

These toxic herbs also provide an important illustration of the importance of 'dose', i.e. the amount consumed. All of these plants contain other alkaloids such as scopolamine and hyoscyamine which are potentially highly toxic, but which have anaesthetic and narcotic properties in very low doses. In contrast, the alkaloid atropine by itself tends to be more stimulatory in its effect.

The stimulatory effect of atropine stems from its role as an antagonist of the neurotransmitter acetylcholine which is an integral part of the parasympathetic nervous system (PNS)[1]. The PNS is responsible for controlling the activity of heart muscle and the smooth muscles of the digestive system. During its normal function acetylcholine attaches to its specific

---

[1] The parasympathetic nervous system controls subconscious bodily functions. In contrast, the sympathetic nervous controls conscious functions such as deliberate muscular activity e.g. locomotion.

acetylcholine receptor (AChR) triggering stimulation of the digestive system and inducing a simultaneous relaxing effect on heart muscle. Atropine reverses this effect by mimicking acetylcholine and attaching to the acetylcholine receptor (AChR) without triggering the calming effect that is characteristic of acetylcholine. This results in the opposite of the so-called 'rest and digest' syndrome that is characteristic of acetylcholine. This stimulatory property has led to the use of atropine as a heart stimulant in some cases of bradycardia (slow heart rate) or heart failure.

The use of deadly nightshade as a narcotic agent goes back several hundred years, with applications ranging from sedation to the induction of a trance-like state. As early as the 15th century it was used as a cosmetic, acting to enhance the eye beauty of 'high society' ladies of the Italian Renaissance by dilating the pupils, hence the species name *Bella donna* (pretty woman). This use of atropine has also been ascribed to the Egyptian queen Cleopatra (69-30 BCE) along with many other tales of her beauty and influence. This pharmaceutical property is reflected in the use of atropine in dilating pupils during eye operations and its current application in the treatment of myopia.

Cyanide. The Deadliest Dark Blue.

Cyanides are found in a few plants such as sorghums and almonds, where the hydrous form, hydrocyanic acid, is responsible for the characteristic 'bitter almonds' flavour associated with these chemicals. Although the cyanide radical $C\equiv N$ is highly toxic, plant cyanides very rarely reach toxic levels through accidental ingestion, and even then, only under exceptional circumstances. Occasional cyanide poisoning results from the consumption of seeds from fruit such as apples, although a toxic dose of >200 mg Kg^{-1} is only be achieved through the ingestion of as many as forty seeds, which would have to be chewed as opposed to being swallowed whole. However, only a small proportion of cyanide sources involve biota.

The toxicity of cyanide is primarily related to its role as an inhibitor of cytochrome C oxidase, an enzyme that plays a major role the electron transfer system which is an integral part of cellular respiration. Inhibition of this system impedes the utilization of oxygen, essentially starving cells of oxygen and inhibiting the ATP[2] synthesis pathway that is driven by the respiratory process. Symptoms of this cellular asphyxiation include dizziness, shortness of breath and subsequent respiratory failure and death.

The history of cyanide as a poison can be traced back to 1704 with a series of experiments by the German painter Heinrich Diesbach. Diesbach was attempting a recipe for a red pigment but, instead, ended up with a deep blue iron cyanide compound that he named Berlin Blue, which eventually came to be known as Prussian Blue. Nearly 80 years later, in 1782 hydrogen cyanide was first prepared from Prussian Blue by the Swedish chemist, Carl Wilhelm Scheele. In its soluble form it is known as prussic acid, named after its Prussian blue source. The word cyan, used to describe the core radical in hydrogen cyanide is another reference to its derivation from Prussian blue, echoing the Greek word '*Kyanos*' for dark blue going back to ancient times.

The extremely toxic negatively charged cyanide ion $C{\equiv}N^-$ is the active ingredient in hydrogen cyanide gas ($H{-}C{\equiv}N$). Sodium ($NaCN$) and potassium (KCN) cyanide salts in their pure forms are highly toxic, tasteless white powders with a

---

[2] Adenosine Triphosphate (ATP) is a nucleoside triphosphate compound responsible for energy production in all living cells.

notorious history as deliberate poisons, chosen for their lack of taste and early difficulties in its forensic chemical detection.

The use of cyanide as a poison gas in the first world war was eventually abandoned because of its lighter-than-air property leading to its rapid dispersal and ineffective deployment in a combat zone. As a poison gas deployed in field combat, it was replaced by phosgene, another manufactured compound produced by passing a mixture of carbon monoxide and chlorine through activated charcoal. Other deadly candidates for trench warfare were a closely related group of compounds, collectively known as mustard gases, although they were more accurately described as fine mists of liquid droplets. These liquid suspensions were characterised by their eponymus mustard-like smell. A notable example of a mustard gas constituent was the sulphurous compound bis(2-chloroethyl) sulphide, $S(CH_2-CH_2Cl)_2$. This was just one of a family of so-called blister agents, a term that accurately describes their effect on human skin and membranes on exposure.

Despite the abandonment of cyanide as a poison gas in field warfare, the use of hydrogen cyanide continued into the Second World War when, in 1942, it became the primary killing agent of the gas Zyklon B in Nazi concentration camps. As a postscript to that dark chapter in human history hydrogen cyanide was also the chosen method of suicide for many of the Nazi leaders of that era, including Erwin Rommel, Adolf Hitler's wife Eva Braun, Heinrich Himmler and Hermann Göering. Another wartime use of cyanide for mass killing was the massacre of between 3000 and 5000 Kurdish people in Halabja, Kurdistan in March 1988 towards the end of the Iran-Iraq war.

In the post World War II era numerous suicides and murders have been attributed to cyanide poisoning. Perhaps the most notorious mass murder-suicide was recorded at Jonestown,

Guyana, in 1978 where more than 900 people belonging to the "People's Temple" cult died by drinking potassium cyanide. Apart from mass killings associated with military, cult or terrorist activity, deliberate cyanide poisoning of individuals or small groups continues to be part of the story of this notorious toxin. However, while illicit financial gain is often behind such murders, motives are sometimes difficult to find. The deaths of six people at the Grand Hyatt Erewan Hotel in Bangkok in 2024 were easily ascribed to cyanide-laced tea, found in the hotel room along with the bodies. However, while financial gain was suspected as part of a hidden agenda behind these killings, the full story may never be known as the presumed perpetrator was apparently among the deceased. A more impersonal 'motive' has been attached to the so-called Tylenol killings of 1982, when a series of seven deaths were traced to a Tylenol pill-container that been purchased in good faith from a pharmacy. Tampering had involved the replacement of the active ingredient of the Tylenol capsules in a specific commercial pill container with cyanide powder. The killings seemed "untargeted", insofar as the victims were random users of the product. The incident resulted in the largest ever mass-recall of a proprietary medicinal brand, numbering over thirty million pill-containers that were pulled from pharmacy shelves. There was an attempt to extort money from Tylenol manufacturers, Johnson & Johnson, threatening further tampering if ransom money was not paid. However, it turned out that when the extortionist was finally identified, they were found to have no connection to the poisoner, who remains unidentified.

Apart from their notorious history as poisons, sodium and potassium salts of hydrogen cyanide have a variety of industrial uses and are currently produced in large quantity in the plastics industry where they are used to produce polymers. They are also used in the mining industry as leaching agents enabling the extraction gold and silver from their ores and act as solubilisers in the electroplating of these metals.

Bearing in mind the history of cyanide as a gaseous poison it is, perhaps, not surprising that hydrogen cyanide has also been employed as an effective fumigant in the food processing industry where it mimics its insecticidal role in plants that are a natural source of cyanide compounds. Plant cyanogens act as inhibitors of cellular respiration and, like many natural toxins, perform a defensive role, protecting plants from attack by herbivores. In response, organisms from numerous herbivorous species are able to counter the action of plant cyanide compounds through a detoxification process involving the enzyme thiosulfate: cyanide sulfotransferase (TST), also named rhodanese. It is therefore clear that part of the rich, yet often deadly history of cyanides can be traced to their role in the biological warfare that is characteristic of so many natural toxins.

Arsenic, The King of Poisons.

Arsenic is a metalloid widely found in the earth's crust, including volcano lava and associated gases which are an abundant source of this element.

Mineral forms of arsenic have been known for over 2,500 years. Aristotle's mention of the mineral 'sandarache' as early as the 4th century BCE is now believed to refer to a sulphide of arsenic known as realgar. This substance, later known as arhenicum was known as a poison in ancient Greece although was also credited with therapeutic effects including as a treatment for syphilis and other diseases. In some quarters it even achieved a reputation as an aphrodisiac, although it's not unreasonable to suggest that this might be linked to arsenic's curative properties. After all, relief from venereal disease could be seen as a boost to a subject's sex-life!

However, it is as a poison that arsenic has gained unparalleled notoriety, earning the title "The King of Poisons". This reputation is largely based on the fact that, in early Greco-Roman times and for centuries afterwards, arsenic poisoning was indistinguishable from many other naturally acquired illnesses and its tasteless presence in food and drink proved impossible to detect. Its toxicity was also cumulative, so a surreptitious series of sublethal doses over time could lead to a fatal outcome disguised as a chronic illness. Chronic symptoms included nausea, progressive weakness, confusion and eventual paralysis.

Some historians have twisted the title "King of Poisons" around to read the "Poison of Kings", a reference to the fact that arsenic was sometimes considered as a means of removing a monarch, Head of State or other important personage through its introduction into their diet.

Such was the popularity of arsenic as a means of eliminating opponents for political and financial gain that in 82 BC Roman dictator Lucius Cornelius Sulla issued a decree against the use of arsenic as a poisoning agent. Nevertheless, it seems that the law was unsuccessful in completely removing arsenic from the scene. In the first century AD, Discorides, a Greek physician at the court of the Roman Emperor Nero, makes reference to arsenic as a poison, and in the 5th century the Greek historian and philosopher Olympiodorus of Thebes provides a recipe for the preparation of white arsenic from arsenic sulfide. We are unaware of his motive for publishing a "how to" on the preparation of the deadly yet tasteless compound white arsenic, otherwise known as arsenic trioxide, (ATO). However, he would have been aware of the reputation of ATO as both a poison and medicine and it seems unlikely that this publication on chemical preparation was designed simply to manufacture a paint pigment!

Nevertheless, there remains a good deal of confusion over the exact form and identity of this element as used in ancient Greece and Rome. This uncertainty continues up to and including the time of the German philosopher and alchemist Albertus Magnus who has been accredited with the "discovery" of arsenic around the year 1250. Again, we are confronted by a claim of discovery without any hard evidence of its exact chemical identity. In this regard it is worth remembering that Albertus was a keen follower of the Greek philosopher Aristotle. His revelations about arsenic may, therefore, simply be a reaffirmation of the existence of a toxic mineral known since ancient times through the writings of Aristotle and his subsequent students.

The first person accredited with the identity of arsenic as an element was the German physician and pharmacologist Johann Schroeder, who, in 1649 isolated elemental arsenic from its oxide. Schroeder's interest in arsenic followed from the teachings of the 16th century Swiss-German alchemist and philosopher Theophrastus Phillipus Aureolus Bombastus von Hohenheim, universally remembered by the pseudonym Paracelsus. Arsenic was one of several chemicals cited by Paracelsus as having therapeutic benefits in sublethal amounts but causing sickness or death at higher doses. In his most famous quote Paracelsus expands on this with the statement:

"*All things are poisons, for there is nothing without poisonous qualities. It is only the dose which makes a thing poisonous*".

Paracelsus, who lived from 1493-1541, is widely referred to as the Father of Pharmacology. Although a keen student of medicine and chemistry he lived at a time long before the isolation and identification of specific chemical compounds and their widespread development for medical and pharmaceutical use. While familiar with the medical properties of natural products of his era, many of his teachings relate to alchemy, often with strong spiritual connotations. The

11

relationship between chemical intake and the occult and spirituality was well-known in the Renaissance period and the notion of 'dose' would have been an important concept. It still is. Some common chemicals, like water and iron, are necessary for life but can be toxic at high 'doses', even water! Other chemicals are so dangerous they are simply considered poisons. However, many poisons have therapeutic uses in sublethal doses, yet a few have been favoured in the commission of murders, suicides and, as we shall see, even executions!

The story of arsenic poisoning continues through the Middle Ages into the 16th century where, in Renaissance Italy, we find the notorious Borgia family intent on creating a family dynasty fueled by the elimination of political and religious opponents who stood in their way. Their principal weapon for doing away with competition was arsenic added to food and wine and served to guests in parties and orgies that were a feature of their decadent lifestyle. Dozens of poisonings were carried out in this way using various arsenic compounds as well as organic additives such as strychnine. An arsenic preparation reputedly favoured by the Borgias was Cantarella, a pleasant tasting white powder notorious for its slow creeping toxicity often mistaken for sickness symptoms associated with other ailments.

Although many of the religious hierarchy numbered among the enemies of the Borgias, the family itself produced two popes; Alfons de Borja, who ruled as Pope Callixtus III from 1455 until 1458, and Rodrigo Lanzol Borgia, who became Pope Alexander VI between 1492–1503. The rule of Pope Alexander VI marked a particularly dissolute period during which the Borgias' activities comprised adultery, incest, bribery, theft and illicit sales of church favours and valuables. These crimes were in addition to many 'murders by arsenic' perpetrated by Alexander. Partners in crime included

Alexander's son Cesare and Cesare's half sister Lucretia Borgia. Some have expressed the opinion that Alexander VI received his 'just deserts' when he became ill by accidentally drinking some of his own arsenic-dosed wine. However, his death in 1503 seemingly resulted from malaria rather than arsenic. Interestingly, not all who came after him judged him badly. His successors as heads of the Catholic Church Sixtus V (1585-1590) and Urban VIII (1623-1644) were especially complimentary ranking Alexander VI as the most outstanding pope since St. Peter!

Perhaps the most prolific poisoner of the post-Renaissance era was the professional assassin Guilia Toffana who, in Rome between 1633 and 1651, is thought to have hastened the journeys of more than 600 men into the afterlife at the hands of their wives and partners. Toffana was responsible for the preparation of arsenic-laced cosmetics with full instructions for their use. Such was her fame that her favourite poison of choice was actually named after her. *Aqua Toffana* was marketed as an oil or perfume but proved deadly when slipped into the food and drink of potential victim or victims. Often the poisoning was carried out in the form of serial droplets over an extended period of time, taking advantage of the cumulative toxic action of arsenic. Guilia Toffana is reported to have been executed for her crimes in 1659 along with some of her accomplices. Hieronymus Spara, another professional arsenic poisoner linked to Toffana is also reported to have been hanged at this time for helping young women to become wealthy young widows.

The possible role of *Aqua Toffana* in the death of the Austrian composer Wolfgang Amadeus Mozart remains highly speculative and seems to be a theory engendered by a fearful statement by Mozart himself close to his untimely death in 1791 at the age of 35. The involvement of arsenic in his death is a topic of strong debate particularly considering Mozart's several genuine illnesses. With the hindsight of history, the most likely explanation for his death seems to be a streptococcal infection leading to kidney disease. Rumours of the involvement of arsenic in his death remain a topic of conjecture. He lived at a time when arsenic was reputed to have a broad range of remedial properties, and it would not have been difficult for him to have obtained arsenic compounds for the treatment of a broad range of ailments. However, more recent evidence suggests that Mozart's own choice for self-medication is more likely to have been antimony, a metal sharing with arsenic a reputation for remedial properties but also some similar toxic characteristics.

In 1786, the British physician Thomas Fowler introduced an arsenic-based, cure-all tonic, appropriately known as Fowler's solution. The solution, which contained potassium arsenite ($KAsO_2$) as the active ingredient, was initially advertised as a treatment for malaria, although its purported remedial powers rapidly expanded to numerous disorders, including psoriasis, leukemia and syphilis. Charles Darwin was a devotee of Fowler's Solution, which he took to treat eczema among other ailments. Fowler's Solution remained a popular remedy and tonic for several years   Unfortunately, it became increasingly apparent by the early 20th century that treated patients were found to have a significantly higher risk of developing cancer, particularly at the point of application of the tonic.

A common thread in tracing the history of arsenic as a poison over millennia has been the difficulty in differentiating specific evidence of its toxicity from the symptoms of numerous other illnesses. As we have seen, without definitive chemical evidence, numerous murders involving arsenic probably went undetected. This came to an end in 1836 when the English chemist James Marsh perfected a sensitive means of detecting the presence of arsenic. The Marsh Test, as it came to be known became a proven method for tracking arsenic in victim's bodies as well as the source of the poison in food and drink. Test results often provided critical evidence in establishing the identity of the poisoner. Such was the case in a series of unexplained deaths in a boarding house in Windsor Connecticut U.S.A. between 1907 and 1917. The proprietors of the boarding house, Amy Archer-Gilligan and her husband specialised in the care of elderly residents in exchange for a fee and, often the bequeathment of other assets at the end of the residents' lifetime. Suspicions arose when, following securement of this fee, the average lifetimes of the residents at the boarding house became shorter and shorter, with the death toll rising to over sixty between 1910 and 1916. The deceased eventually included Amy's two husbands James Archer and Michael Gilligan, with the latter marriage lasting only three months. Many of the deaths seemed to result from bilious attacks as well as heart failure although, considering the advanced age of many of the deceased, the number of deaths due to arsenic poisoning remains uncertain. Forensic diagnoses were further complicated by the role of arsenic as a component of embalming fluid. However, Amy's conviction in 1917 was sealed by the discovery, using the Marsh Test, of enough arsenic to kill a dozen or more people in the stomachs of five of the bodies; this, and evidence of her multiple purchases of arsenic from a pharmacy "for pest eradication".

Although this was only one of several cases of arsenic poisoning on record it gained a degree of notoriety as the inspiration for Joseph Kesselring's 1941 play "Arsenic and Old Lace" featuring two 'sweet little old ladies', Abby and Martha, who hastened the demise of lonely male lodgers in their guesthouse with poisoned wine before burying them in the cellar. This, in turn led to the 1944 Frank Capra film of the same name starring Cary Grant. The 'model' for Abby and Martha, Amy Archer-Gilligan, was spared the death penalty following a successful insanity plea and died of old age in an asylum in 1962.

Despite its long history as a poison, the therapeutic use of arsenic continues to be of significant current interest and we are again reminded of the teachings of Paracelsus who cited arsenic as an example of a poison with medicinal properties at sub-lethal doses. This phenomenon is well illustrated by looking back through the lens of history, beginning with the *Shen Nong Ben Cao Jing*, the first Traditional Book of Chinese medicine dating from around 200 BC. The book recommends realgar (arsenic sulphide) as a remedy for skin conditions such as boils, psoriasis, insect bites and carbuncles as well as a treatment for many internal maladies including malaria, intestinal parasitosis and epilepsy. Among the curative properties attributed to realgar were "detoxification" and "cooling of the blood", i.e. fever reduction. For internal use the preferred method of administering realgar was in the form of Niuhuang Jiedu tablet (NJT), a traditional Chinese medication retained to this day.

A recent development in the clinical use of arsenic with its roots in traditional Chinese medicine has been the demonstration of arsenic compounds as potential cancer treatments. Beginning in the 1970s both arsenic sulphide (realgar) and arsenic trioxide (ATO) have undergone extensive testing as treatments for several different types of cancer including acute promyelocytic leukemia (APL), malignant lymphoma and chronic myeloid leukemia (CML). Although some remedial success has been claimed for both chemicals, questions remain over contradictory evidence of arsenic carcinogenicity on one hand and the anti-cancer properties of realgar and ATO on the other hand. Obviously, the mechanisms underlying these contradictions will need to be fully understood before risks associated with the use of arsenic as a cancer treatment can be satisfactorily eliminated.

Arsenic exists primarily in mineral form where it forms numerous inorganic compounds, often with metallic elements.

Iron, cobalt, lead and nickel arsenicals are major industrial sources of arsenic which have a wide range of uses. These include its employment as a strengthening agent in a lead alloy used the battery and semiconductor manufacture. Arsenic has also been widely used in a broad spectrum of pesticides although this has been curtailed in recent years following increasing concerns over its toxicity to humans and domestic animals.

The inclusion here of arsenic as a Natural Toxin extends the use of the word 'natural' to chemicals not directly originating from the biosphere. These include toxic metallic elements, such as lead, cadmium and mercury, that are usually extracted from minerals rather than biological sources, but whose uptake by humans and other animals often results from human activities such as mining, smelting and various manufacturing industries. Accounts of the history of trace metals and their toxic legacy appear later in this book.

Clearly the forementioned 'top three' toxins, Atropine, Cyanide and Arsenic, have little in common apart from their notoriety as poisons and defy any convenient categorisation relating to their source, structure or Mode of Action. Yet they all contribute to a rich history dealing with a vast range of Natural Toxins that have shaped our history and continue to play important roles in the natural and human worlds.

The Evolution of Toxicology.

Before embarking on a detailed account of the many different natural toxins, their sources and effects, it is useful to take a look at the history of the methods that have been developed and refined over many hundreds of years to detect and measure the biological effect of these compounds. This knowledge has been critical in assessing their specific properties and, importantly, identifying the line between harmful and beneficial effects.

The study of natural toxins and poisons can be traced back to ancient times. The writings of alchemists and philosophers such as Aristotle provided information on the sources and preparation of natural chemical substances that possessed medicinal and narcotic properties but also produced harmful effects depending on the amount consumed and method of ingestion. Much of this early usage was empirical in nature in the sense that, in the absence of any chemical knowledge or reliable analytical techniques, estimates of the efficacy and risk attached to these compounds were often a matter of 'trial and error'.

This situation began to change with the development of toxicology as a scientific discipline devoted to the understanding of how organisms become exposed to potentially toxic chemicals and the nature of the toxic response. For this we are once again indebted to Paracelsus, who was an anti-establishment figure not afraid of ruffling feathers. He embarked on a crusade to place the science of toxicology on a sure footing by establishing a reliable relationship between the amount of toxin ingested and its effect.

Bioassays and the Concept of Dose-Response.

Over the centuries since the writings of Paracelsus, the relationship between chemical dose and the resulting biological effect has been further refined into what is now regarded as classic toxicology. In this scientific discipline, biological responses to chemical exposure are typically expressed as reproducible biological endpoints, known as dose-responses, which are obtained from controlled laboratory experiments. These tests, known as bioassays, are designed to measure the toxicity of specific chemicals or groups of chemicals, most of which have only been identified over the last two centuries.

Bioassays are essentially experiments that are designed to follow one or more biological endpoints or outcomes following exposure of test organisms to a prescribed range of chemical concentrations over a specified period. Short-term assays known as acute toxicity tests typically follow "all-or-nothing" endpoints such as mortality i.e. live/dead, whereas longer-term, 'chronic' tests tend to focus on more subtle sub-lethal end-point characteristics such as growth or behavioural activity. Many bioassays take advantage of our newfound knowledge of molecular biology and biochemistry and are used to track physiological and biochemical responses in biota exposed to 'foreign chemicals'. Chemical responses in biota may result from exposure to other organisms that may be the source of natural toxins but may also be influenced by changes in the physical environment, such as those brought about by climate change. These are topics examined further in the context of the role of natural toxins in interspecies competition.

The length of a toxicity bioassay may be determined by the size and taxonomy of a test organism. For example, laboratory test species may range from rodents such as mice and guinea pigs at the upper end of the size range to microscopic organisms including bacteria.

End-point observations for acute aquatic bioassays using standard test organisms are typically made at 24h, 48h and 96h intervals following initial chemical exposure. However, for larger laboratory test animals such as rodents, 'acute' measurements of adverse symptoms may extend to a matter of a week or more depending on the normal lifespan of the test organism. What constitutes an acute time span for one species may represent most, or all, of a life cycle for shorter-lived species. In chronic assays biological endpoints might extend to more than one generation and may include reproductive performance as determined by number of offspring produced following toxic exposure. For rapidly growing microorganisms such as bacteria, reproductive performance constitutes the sole means of assessing toxicity, which is typically measured in terms of 'doubling time'. This is the time taken for the bacterial test population to double in numbers following exposure to toxic stress or some other form of environmental change. In such bioassays more stressful conditions will result in a longer doubling time or even 'negative growth' i.e. a reduction in bacterial numbers.

The 1920s saw the introduction of a method for determining toxicity based on the chemical dose required to produce a specific measured response. This experimentally derived 'dose-response' therefore provided a means of ranking chemicals according to their toxicity.

Laboratory assays are usually conducted under uniform conditions, on standardised, replicated sets of test organisms. A typical assay might consist of several groups of organisms simultaneously exposed to a range of chemical concentrations. Exposure concentrations typically cover a logarithmic or semi-logarithmic series and are usually preceded by range-finding assays designed to define and refine the full response range. In graphical form, the end-product is a dose-response curve covering the complete range of responses from zero to full (100%). Usually, bioassays contain four or more replicates exposed to a range of concentrations of the test chemical, including 'control treatments' wherein no chemical is added. Essentially 'controls' represent "zero dose" and are designed to account for any stress that might result from ambient conditions such as experimental temperature, illumination and any other extraneous factors. Standard test organisms may range from laboratory rats and mice to small aquatic organisms such as larval fish and invertebrates. However, depending on the nature of the toxin to be tested test organisms may also encompass microscopic and sub-microscopic organisms and cultured cells of various kinds.

In Vitro Testing and High-Throughput Sampling.

The post-industrial era reflected a more streamlined world looking to capitalise on the great strides that had been in engineering and the manufacture of new chemical compounds for a variety of industries. In the medical world new

---

Determination of Dose-Response.

The most common expression of dose-response is the LC_{50}. The LC_{50} is defined as the "Lethal Concentration", hence LC, that results in a 50% mortality of the test population (i.e. the total number of test organisms used) following a specified exposure time. So, for example there might be a 24-hour LC_{50} or a 96-hour LC_{50}. Although 50% mortality is the conventional endpoint normally chosen, sometimes a different level of mortality such as 25% might be selected as the endpoint, in which case the measure becomes LC_{25}. Sublethal endpoints are also measured in this way. Where a sub-lethal effect is measured, the term 'L' for Lethality is substituted by a generic 'E' for 'Effect', no matter what that effect might be. For example, the chemical concentration resulting in a 50% reduction in growth or the concentration causing a halving of the number of offspring are both expressed as EC_{50}. A hundred years after their first use, LC_{50} and EC_{50} remain the most common measures of chemical toxicity and still form the basis for assessing the biological effects of natural toxins and other chemical compounds on individual organisms.

---

pharmaceuticals, and antibiotics have benefitted from many decades of rigorous testing to establish safe and effective doses for these compounds, many of which were derived from natural toxins. Both acute and chronic toxicity bioassays using laboratory animals such as rodents and rabbits became a dominant feature of clinical toxicity testing in the first half of the 20th century. The use of mammalian test species was justified on the basis that the end-users of these medicines and pharmaceuticals were primarily human subjects and their (mammalian) domestic animals. The term 'guinea pig' has since become synonymous with this testing culture and remains part of the language used to refer to the trial of an unknown substance or procedure. Although bioassays continue to form the basis for much of the fundamental toxicological data currently available on natural toxins, several changes to testing procedures now exist. As early as the 1950s misgivings over perceived cruelty to mammalian test animals prompted an increasing number of prohibitions. For example, several U.S. states currently ban the use of cosmetics that have been tested on mammals. These, and several similar restrictions have led to many refinements in test procedures in recent decades including the increased use of tissue and cell culture, known as *in vitro* techniques as surrogates for live animal *(in vivo)* testing. As humans are usually the end-users for most pharmaceuticals derived from natural toxins, the use of human cell lines has bolstered the case for this approach.

In the modern pharmaceutical industry, however, justification for the employment of *in vitro* techniques for testing natural toxins for their potential as medicines and drugs goes well beyond concerns for animal welfare. The advent of high-throughput sampling and associated assays in the late 1980s and early 1990s paved the way for a quantum leap in the scale of detection and identification of causal 'events' triggered by natural toxins and their derivatives. Using this technology, a massive array of chemical variants and doses can be exposed to a matrix of biological and chemical substrates ranging from tissue and cell cultures to macro-chemicals such as proteins and nucleic acids.

Much of the success of this approach has to do with sampling plates and automated nano-[3] sampling techniques capable of processing of as many as a million tiny samples an hour. Signals generated by high throughput sampling (HTS) are not necessarily directly toxic themselves, but many are capable of providing clues signifying the activation and modulation of important biochemical pathways. For example, a change in the light absorbance pattern can often indicate chemical binding to a nucleic acid or a protein which may function as an enzyme. These, and other endpoints such as antibody formation and gene activation provide important clues, often described as 'hits', that lead to a more detailed toxicological examination and the selection of compounds and doses that maximise pharmacological potential. The involvement of genetic building blocks such as nucleic acids adds another dimension to the screening process, opening the door to the use of 'genomics' as an extension of HTS.

Genomics.

Genomics, in the form of gene mapping and editing, can be employed to develop and manipulate the pharmacological and medicinal potential of chemicals according to their structure and dose. Gene mapping identifies unique sequences of DNA nucleotides that code for specific chemical groups via RNA templates. These are manipulated by cutting and repair techniques to edit the DNA to modify or create novel chemical structure(s). These techniques may be employed to modify the genetic make-up (genome) of an organism to select for specific biochemical or physiological characteristics. In certain instances, these changes may be seen as correcting a

---

[3] Nano as a prefix means one billionth or 0.000000001. Informally, nanotechnology is sometimes used to describe very tiny volumes.

deficiency.

An example of this application is the use of a remarkable piece of gene editing technology known as Clustered Regularly Interspaced Short Palindromic Repeats, shortened to CRISPR. At this point we need to forgive geneticists for their arcane use of language, and to remember their extraordinary facility for recognising coded genetic sequences in the same way that expert linguists might identify alphabetic combinations. Repeating nucleotide sequences in DNA from the bacterium *Escherichia coli* were recognised in 1987 by the Japanese geneticist Yoshizumi Ishino as a form of antiviral immune response. The gene editing application of this technology won the 2020 Nobel Prise for Professors Jennifer Doudna and Emanuelle Charpentier who recognised the potential of the *cas* enzyme and its ability to snip and repair the double-stranded DNA molecule at specific locations following the incorporation of viral DNA. The point(s) of DNA excision and recombination are dictated by the site of incorporation of viral DNA in relation to the repeated (CRISPR) sequences. The technique provides a means of removing mutations responsible for inherited diseases, thereby silencing their effect.

An example is the ablation using CRISPR cas-9 of the rhodopsin gene carrying a mutation responsible for *retinitis pigmentosa*, a hereditary disorder characterised by degeneration of the retina and loss of sight. Laboratory experiments using rodents have demonstrated the effectiveness of gene editing in mitigating the development of retinal degeneration in affected animals.

As a postscript to this, assuming this disease was responsible for the affliction of the "three blind mice" of nursery rhyme fame, imagine the part that CRISPR cas-9 might have played in rescuing these unfortunate rodents from their fate!

'Three blind mice, (See how they run.)'
Three mice, victims of Retinitis,
in mortal danger, fail to notice the knife
In the hand of the farmer's wife
A detail that could well entail them
prematurely losing their life
Or at the very least being de-tailed

But this time no fear of losing their life
At the hands of the farmer's knife-wielding wife.
No problem with losing their tails this time,
Because of their treatment with CRISPR Cas-9.
The reason the CRISPR Cas gets all the credit
Relates to its predisposition to edit
The gene that is causing the blindness
Thus dealing the mice the remarkable kindness
Of restoring sight and their optic acuity.
Perfect vision regained in complete perpetuity.

Now that the animals see where they're going
It's clear that the peril of toing and froing
Near someone possessing a large carving knife
Presents a real danger to limb and to life.
Far better to make a run for the door
Or anywhere that will serve to ensure

A happy conclusion to this little tale

With no loss of life or disjoining of tails.

The moral of this story means

The answer may be in our genes.

And that we may stand to prosper, as

We learn more about CRISPR-cas.

In many respects, gene editing in association with HTS can generate such data on as many as 100,000 chemicals per day. Clearly the title 'high-throughput screening' is well earned!

A feature of HTS is the sheer volume of information generated within a short period of time, often necessitating the use of computer programmes capable of handling the enormous mass of data produced. Using computerised algorithms it is now possible to identify structural patterns in chemical sub-groups or molecular configurations that are responsible for specific biological properties. Patterns may range from specific nucleic acid sequences that form the templates for specific chemical groups to characteristics associated with the chemical subgroups themselves. The linkage of genetic information to the synthesis of specific chemical compounds may be a critical factor in developing effective drugs and medicines from natural toxin sources. In this respect a major incentive for the genetic manipulation of a chemical structure is the selection of a particular chemical sub-group that is related to a specific biological activity. As examples, hydrocarbons such as terpenes and nitrogen-containing cyclic compounds classified as alkaloids have an enormous range of biological activities many of which have pharmaceutical and other useful properties. Gene-editing techniques are routinely used to favour the selection of chemical groups that enhance the activity of specific bioactive substances such as medicines, pharmaceuticals and pesticides.

Much of the material involving current and projected screening processes for specific biochemical transformations and pathways is beyond the scope of this book. After all, we are dealing here primarily with the history of natural toxins, not their future. Nevertheless, the narrative provides an illustration of how quickly modern technologies and ideas become history. Despite the recent evolution of sophisticated screening techniques that help define the link between chemical structure and biological effect, the history of this relationship goes back more than 150 years.

Structure Activity Relationships (SAR).

Probably the earliest definition of what is now known as the Structure Activity Relationship (SAR) can be traced back to the discovery of the alkaloid physostigmine in 1846 by Scottish toxicologist and physician Robert Christison. Physostigmine (Figure 2.) is a natural toxin derived from the Calabar bean which is the seed of the leguminous plant (*Physostigma venenosum*) native to African countries such as Kenya and Nigeria where it had achieved notoriety as a means of execution for witchcraft and other 'crimes'. Its role in these 'trials' could be described as judge, jury and executioner.

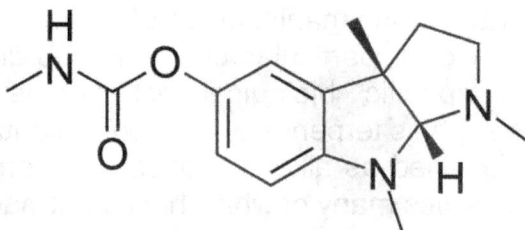

FIGURE 2. Physostigmine.

If the accused ingested the beans without dying, they were assumed to be innocent and released. Nevertheless, it seems reasonable to question the role of the mandatory dose in this context, i.e. how many toxic beans assured a guilty verdict?! An indigenous name for the Calabar bean, 'e-ser-e', gives rise to the often-used alternative name for physostigmine which is eserine. Another source of physostigmine is the fruit of the Manchineel tree found in subtropical South America and the southern states of North America. Little more needs to be said beyond the title 'Tree of Death' given to this plant.

More than fifteen years after Christison's publication of the sources and toxic properties of physostigmine it was the pioneering study of the compound by A. Crum Brown and Thomas R. Frazer in the 1860s that formed the basis for branch of toxicology we know today as Structure Activity Relationship (SAR).

The seminal work of Brown and Frazer on the alkaloids physostigmine and atropine demonstrated a clear link between the chemical structures of compounds and their biological properties:

"There can be no reasonable doubt that a relation exists between the physiological action of a substance and its chemical composition----" (Brown & Frazer 1868)

This and later work elucidated the role of physostigmine as an anticholinesterase inhibitor that acts as a stimulant of the parasympathetic nervous system (PNS). The stimulatory effect of physostigmine on the neurotransmitter acetylcholine gave the work importance in establishing SARs as critical tools in the development of new pharmaceuticals and medicines. In another aspect of this research, they observed that physostigmine and atropine displayed similar biological activity when applied individually yet behaved antagonistically when both were present. One conclusion drawn from this research was that compounds with similar structural characteristics can act on biological systems in a similar manner, yet when applied together they may act competitively. However, even in this early work it was clear that the degree of competition was complex and dose dependent, i.e. the antagonistic effect of one chemical on another could be lessened or reversed depending on their absolute or relative concentration(s). Now, more than 150 years later, it is understood that Structure Activity Relationships tend to work in our favour when it comes to the development and refinement of new drugs and medicines. However, even with structurally similar compounds, variation in biological effects can hinge on small differences in chemical sub-group position and configuration. This is a theme further discussed when considering the additive or antagonistic role of biologically active toxins in their natural environment and their adoption and refinement for human use.

Much of the history of natural toxins is bound up with their discovery, isolation and potential for industrial and medical application. However, many chemical properties that predispose certain compounds for human adoption and use are also those that benefit the biota that produce those compounds in the natural environment. As we have seen from the study of SARs, the biological activity of specific natural toxins often results from a specific chemical sub-group. Different compounds ingested or applied at the same time can interact competitively depending on their chemical structure and may elicit either antagonistic or additive biological responses. Of course, the form of these biological effects will also differ depending on the nature of biota that are exposed to chemicals either singly or in combination. Variability in the response of different plant and animal species to natural toxins is a critical factor in determining the role these compounds play in ecosystems.

When considering the role of toxins in the natural environment it is important to bear in mind that chemical interactions with biota do not occur in isolation. Unlike the case of a single organism exposed to a chemical or a pair of chemicals, chemically induced biological effects in the ecosystem as a whole may be detected at all levels of biological organisation from molecules to ecosystems. Generally speaking, higher order changes in biological communities and populations result from cumulative effects at the molecular, tissue and species levels. This hierarchical approach is reflected in the language used to characterise chemical toxicity in the natural environment, although the terminology used can be inconsistent and often requires further definition.

For example the word 'toxicity' tends not to be applied to adverse effects of chemicals at the population level and is more commonly used to describe the negative effects of specific chemicals to individual organisms. However the word 'Ecotoxicology' implies an emphasis on chemical effects at the level of ecosystems and their components (Campbell et al. 2022).

The Advent of Ecotoxicology.

In food webs indirect biological effects may result from differences in sensitivity to chemical toxins among the species that make up a particular ecosystem. For example, if a predator species or consumer is highly sensitive to a specific natural toxin or other environmental stressor, its elimination or depletion is likely to favour the growth and survival of prey or food species that are more resistant to that toxin or environmental change.

The complex effects of toxic chemicals on different components of food-chain became increasingly evident in the 1960s, partly as a result of chemical spills and other large-scale environmental 'events' which had ramifications throughout specific food-chains and at the ecosystem level.

Prominent among case studies of industrial pollution was the contamination of the Shiranui Sea by mercury from an industrial outfall at the town of Minamata on the island of Kyushu in southern Japan. Although the mercury at the source was mainly mercury sulphate, it was converted by bacteria to the more bioavailable and toxic form, methylmercury (MeHg), which entered the food chain in shellfish and finfish, eventually reaching human consumers. Seafood contamination by MeHg continued through the 1950s and 1960s giving rise to a debilitating and often fatal neurological disorder that was given the name Minamata disease. A feature of MeHg poisoning was the lipid solubility of the compound, resulting in its particular concentration in fatty neurological tissue. This high lipid solubility also resulted in a phenomenon called biomagnification, which was characterised by a cumulative increase in the tissue concentration of mercury in organisms further up the food chain. At the top of the food chain, humans and their domestic animals were, therefore, particularly susceptible to a range of neurological disorders caused by MeHg.

Increasingly, in the 1960s and 1970s the world was becoming faster and more interactive. Global travel was no longer the prerogative of the rich and ruling gentry and all forms of transport expanded massively in volume. Travel consumed energy, and this in turn meant an upsurge in the extraction, refinement and combustion of fossil fuels for energy production. This had widespread consequences. Beginning in the 1960s the world became increasingly familiar with oil spills from tankers, pipelines and drilling platforms. At least 14 major spills made front page news within a few decades. Each incident carried its own set of characteristics and causal factors. Perhaps the first to 'hit the headlines' in a big way was the 'Supertanker' *Torrey Canyon*, which went aground off the southwest coast of England in 1967. Other incidents included oil rig explosions and collapses, sometimes involving loss of lives, and resulting in the release of massive volumes of oil. The largest of these was the *Deepwater Horizon* oil spill in the Gulf of Mexico in 2010 resulting in the release of over 700,000 tonnes of oil forming a slick of about 150,000 square Km. These were just a few among many instances of energy-related events capable of affecting ecosystems on a large scale.

Other pollution effects often associated with energy production proved to be truly global in scale. Oxides of sulphur (SO_x) and nitrogen (NO_x) released during the burning of coal, were identified as the prime causes of acid precipitation, resulting from the dissolution of these gases in rainwater in the form of sulphuric acid (H_2SO_4) and nitric acid (H_2NO_3). Acidification of freshwater bodies became a major environmental concern in the 1970s, with 'Acid Rain' becoming a common topic of conversation. As a talking point this term has been somewhat overshadowed in recent decades by another common phrase, 'Global Warming', which is also related to energy production. The primary cause of global warming is the atmospheric release of increasing levels of carbon dioxide through the combustion of fossil fuels, either for direct energy production to propel vehicles or for the generation of electricity that is fast becoming another means of powering cars and other forms of transportation. In view of the complexities involved with interpreting atmospheric events and their consequences, the term 'Climate Change' is often preferred to Global Warming.

At this point it is reasonable to ask, "How are Acid Rain and Climate Change related to the biological effects of natural toxins?". There are several aspects to this, although all are interrelated. Most concern the biological effects of toxins in their natural environment, as opposed to their application and use by humans which usually happens under relatively controlled conditions. Changes in global temperatures have varying effects on the susceptibility of biota to natural toxins depending to their relative tolerance of the additive stresses of temperature change and chemical exposure. A similar situation relates to the acidification of aquatic habitats where organisms are faced with the potentially additive physiological effects of increased acidity and natural toxin exposure. In some cases, there is the added complication that acid conditions can alter the chemical form of some compounds making them more bioavailable. Where temperature change and acidification occur simultaneously there may be further additive effects associated with the exposure of biota to natural toxins.

Climate-related shifts in the habitat of animal and plant populations may alter exposure patterns to natural toxins depending on the scale and distribution of these chemicals. Often the toxins themselves are associated with biota. For example, the periodic warming of the tropical Eastern Pacific

34

Ocean known as *El Nino* occasionally results in fish kills coastal areas of South America. In March-April 2021, a large bloom of the toxic alga *Heterosigma akashiwo* triggered by the influx or warm Pacific water into the coastal area of southern Chile caused many thousands of salmon deaths in Comau fjord. Here was an example of the direct effect of climate change on the exposure of a fish to the natural toxin. The toxin associated with this highly toxic algal species has been provisionally identified as the powerful neurotoxin brevetoxin or one of its derivatives, although other natural toxins including polysaccharides and powerful oxidising agents have also been implicated. Since the 1980s several climate-related blooms of *H. akashiwo* have resulted in significant economic losses due to wild and farmed fish kills in coastal areas of both the Atlantic and Pacific oceans.

Whether we are talking about large-scale pollution events or the broad influence of environmental conditions on the biological effect of natural toxins it is impossible to fully appreciate the impact on the biosphere without paying attention to the direct and indirect effects on all components of the ecosystem. It is these complexities that gave rise to the relatively recent science of ecotoxicology.

The term 'ecotoxicology' was first used by French toxicologist René Truhaut in 1969, and was defined as the integration of the:

"----toxic effects caused by natural or synthetic pollutants on the constituents of ecosystems, animal (including human), vegetable and microbial ---".

The term covers a broad range of detrimental effects over the full spectrum of biological organisation from individual species to multispecies communities and food webs. In food webs indirect biological effects may result from differential chemical toxicity among taxonomic groups. For example, if a predator species or consumer is highly sensitive to a specific natural toxin or other environmental stressor, its elimination or at least reduction in numbers is likely to favour the growth and survival of prey or food species that are more resistant to that toxin or environmental change.

Throughout the whole biological spectrum from the cellular to multispecies levels there are numerous toxicological endpoints that provide insight into the roles that toxic chemicals play in the natural environment. Many of these endpoints provide a quantitative or semi-quantitative measure of this impact. For instance, complex multispecies interactions at the ecosystem level can form the basis for Environmental Risk Assessments that are commonly used by environmental agencies to estimate the probability of a negative environmental outcome resulting from human activities such as chemical discharge or changes in land use or resource management.

Classical Toxicology.

Other disciplines such as clinical or forensic toxicology are generally seen as branches of classical toxicology that are used to measure the relationship between chemical dose and biological response for specific chemicals or closely related groups of chemicals. Such 'dose-response' values are typically generated obtained from controlled laboratory experiments on a single biological species or group of species following measured chemical exposure or dose. These tests, often referred to as bioassays are designed to obtain toxicity data for specific chemicals or classes of chemicals. Bioassays are designed to follow changes in biological endpoints (e.g. growth or mortality) that are associated with the exposure of test organisms to a prescribed range of chemical concentrations over a specified time period. Short-term assays typically follow acute "all-or-nothing" endpoints such as mortality, whereas longer-term, so-called 'chronic' tests tend to focus on more subtle sub-lethal characteristics such as growth or behavioural traits such as motility or feeding behaviour. Chronic tests also record the development of physiological and biochemical responses to extraneous chemical and biological stresses. The length of a test may be determined by the size, taxonomy and habitat of a test organism. End-point observations for acute aquatic bioassays using standard test organisms are typically made at 24h, 48h and 96h intervals. However, for larger, laboratory test animals such as rodents, 'acute' measurements may extend to a matter of weeks depending on the normal lifespan of the test organism. What constitutes an acute time span for one species may represent most, or all, of a life cycle for shorter-lived species. In chronic assays biological endpoints may extend to more than one generation including reproductive performance as measured by number of offspring.

When it comes to a choice of test organism and specific chemical(s) to be tested, bioassays may consider localised factors such as the need to assess an environmental threat to a key species in a specific community. Other priorities might relate to a particular discharge, spill or other source of pollution and may, therefore, focus on a specific chemical or group of chemicals. In cases where a particular risk to a specific organism has been identified, the time scale of the test may be tailored to the life cycle of that organism, possibly including a record of survival data over more than one generation. In such cases reproductive success might be a critical chronic endpoint. At the microbial end of the biological scale the outcome of a bioassay might be measured as a change in the doubling time of a specific microorganism as a function of chemical exposure. As the name suggests, the doubling time of a bacterial culture is the time taken for bacterial cells to double in number. Doubling time is, therefore, inversely related to the toxicity of the chemical medium under investigation, i.e. a longer doubling time indicates greater toxicity.

Laboratory-derived toxicity data can also provide information on the interactive properties of specific components in chemical mixtures, including how specific compounds can significantly modify toxic effects of other chemicals present. The synergistic[4] and antagonistic[5] characteristics of naturally occurring toxins have a critically important bearing on their biological effects and their function in the natural environment. This information is also highly useful in cases where natural toxins and their derivatives have been synthesised, or the

---

[4] Synergistic effects are more than additive, i.e. greater than the added effects of constituent toxins; analogous to 1 + 1 = 3

[5] Antagonistic effects are less than additive, less than added effects of constituents; analogous to 2 + 2 = 3

compounds themselves have been adopted for human use as narcotics, medicines and poisons. In addition to assessing the degree and nature of chemical toxicity, such bioassays are critically important in evaluating safe or even beneficial sub-lethal doses.

It should be noted that the term 'bioassay' has been used here to describe a controlled experiment designed to quantify the effect(s) of a chemical or chemical mixture on specific organisms or groups of organisms. Adding the word 'toxicity' signifies a negative outcome and in this sense the term 'toxicity bioassay' is synonymous with 'toxicity assay'. Some texts prefer a narrower meaning for 'bioassay', restricting it to a purely biological measure of change or stress irrespective of any known chemical exposure. In such a case the response a test organism exposed to an environmental medium becomes the sole indicator of any potentially toxic effect, even in the absence of chemical data. As an example, bioassays such as these are used to assess the risk to a potential crop species introduced into new environment that may be of unknown chemical status. A reduction in crop yield can be used as an empirical endpoint that can be used to assess the level of environmental stress. Results of field assays may result in changes to crops that are more resistant to ambient chemical and other environmental stress factors. Yes: Plants can act as test species too! Assays can be carried out in field plots or a laboratory setting using soil or soil leachate from a specific site projected for cultivation. Such tests have been used to assess the suitability of a particular location for the introduction of a crop species in the face of resistance by native plants. Resistance to these 'invasive plant species' (i.e. newly introduced crops) may take the form of natural toxins exuded by the roots of native plants. The toxicity of these compounds is an example of a phenomenon known as Allelopathy whereby plants engage in a form of biological warfare to gain supremacy over potentially competitive neighbours.

Following a review of the methods that have evolved to assess and measure the biological effect of natural toxins we now turn to the identity of these compounds, where to find them and their role(s) in the natural world. The inclusion of arsenic and mercury as a Natural Toxin extends the use of the word 'natural' to include chemicals not directly originating from the biosphere. These include toxic metallic elements, such as lead, cadmium and mercury, that are usually extracted from minerals rather than biological sources, but whose uptake by humans and other animals often results from human activities such as mining, smelting and various manufacturing industries.

Natural Toxins in the Environment.

A more conservative definition of 'Natural Toxins' limits the use of this term to a broad range of organic compounds, mainly but not exclusively originating from biota, that often have pronounced physiological actions on other organisms including humans. They comprise a massive number of chemicals derived from a broad spectrum of organisms, mainly plants. Such compounds can cause a wide range of adverse effects on neighbouring biota following exposure. Toxicity may result from ingestion and inhalation of these compounds or physical contact with the organism that is the source of the toxin. In such cases the toxic chemicals may be contained in an exudate or may be associated with a physical feature such as a barb or sting.

Many natural toxins are considered to be protective in nature, defending their producers from attack or consumption by other organisms or providing a competitive advantage. In the most extreme cases these chemical compounds can cause the rapid death of potential predators and consumers. However, in many cases 'natural toxins' simply confer a noxious smell or taste on potentially edible plant tissues as a means of warding off herbivores. At first sight the roles of natural toxins, such as the causation of predator or consumer mortality on one hand and repulsion/avoidance on the other hand, may be regarded as opposite ends of a complex spectrum of defensive strategies employed by biota to increase their survivability and reproductive potential in the face of the depredations of other organisms or competition from neighbouring biota. The term 'strategy' is used advisedly here, as it implies an active response to a specific stimulation. Even the word 'defence' brings an active component to what has generally been seen as an evolutionary process lasting many millions of years.

The Evolution of Natural Toxins.

The evolutionary process itself has featured the use of dynamic terminology such as 'the survival of the fittest', despite an underlying presumption that this process might be seen as a random series of events shaped and influenced by biotic and other environmental factors over many millennia. In this context, the development of new natural toxins, their structure, biological source and concentration is moulded by the challenges presented by the interaction with other biota and the physical environment.

The resulting chemical compounds are seen as the product of a natural selection process favouring the formation of chemicals that enhance the survival and reproductive performance of the organisms that produce them. Viewed in this way the process can be seen as a means of propagating the genetic apparatus responsible for the production of specific groups of biologically active chemicals, i.e. Natural Toxins. This selection process extends to the production of specific enzymes and other chemical groups capable of transforming the harmless products of metabolism, known as secondary metabolites, into compounds best suited to the defence of their producer. The resultant genetic template ensures that specific natural toxins that can satisfy the defensive needs of specific organisms will be carried into future generations of those organisms. Of course, selective pressures also apply to the generation of counteractive chemicals by competitive organisms that detoxify or in some other way prevent the harmful effect of these toxins.

One way of looking at this is that organisms whose survival has relied on the defensive or predatory role of certain natural toxins may act as 'conduits' or 'repositories' for these chemicals or families of similar compounds in the biosphere. Essentially this represents a form of 'mutual support' whereby the propagation of the toxin "compound X" is aided by the enhanced survival and proliferation of the organism that compound X helps to protect. In other words, 'the chemistry helps the biology' and 'the biology supports the chemistry'. When expanded to include the protection of genetic material responsible for the synthesis of compound X, this reciprocal support system bears some resemblance to the 'Selfish Gene' analogy as proposed by Dawkins (1976), or at least a part of that narrative. Using this analogy, Natural Toxins might be seen as one of several genetically selected components that help to ensure the survival and growth of the source organism in a competitive environment.

Defence or Offense?

So far, reference to natural toxins has been largely confined to a defensive role for these compounds. However, competition may encompass more aggressive deployment of natural toxins by biota as a means of inhibiting the growth and reproduction of neighbouring organisms. Competition may occur for the same resources, including food, water and, for plants, access to sunlight. Plant characteristics such as gamete[6] survival and dispersal, seedling germination, and growth all play a part in the ability of the producer of chemical toxin(s) to thrive and flourish in the face of competition from other biota. In motile organisms such as animals, speed and agility are often critical in feeding and predator avoidance. Yet, the competitive edge may also be enhanced by measures that include the active inhibition of these same properties in competitors. Natural toxins play an important part in many aspects of this competitive process.

---

[6] Gametes are male or female cells, plant or animal that unite to form a "zygote", which develops into a new embryo (offspring).

Allelopathy.

Allelopathy in plants is usually described as a phenomenon whereby biochemicals released by a plant have an inhibitory effect on pollen germination, seedling growth, reproduction, and survival of neighbouring plants that occupy a competitive environmental niche. Sites of action for allelochemicals include cell division, nutrient uptake, pollen germination, photosynthesis, and a variety of enzymic functions.

Occasional references can be found in the literature to positive effects on other plants although allelopathic effects are generally regarded as negative. This is reflected in the word 'allelopathy' which is derived from the Greek 'allelo' (mutual harm) and 'pathy' (suffering). Positive effects of allelochemicals on nearby plants and animals tend to be the exception rather than the rule and generally occur where the neighbouring biota perform some beneficial function such supplying nutrients or protection against common pests or consumers.

Allelopathic compounds comprise many different chemical groups including those responsible for deterring potential consumers and other pests. The primary difference here is that, unlike predator-prey interactions, allelopathy does not generally involve ingestion or even direct contact with target organisms. Although some references have extended the definition of allelopathy to include the deterrence of plant pests, the term is most commonly applied to chemically induced hindrance of competing plants in close proximity to one another. Inhibition of the growth, reproduction and survival of nearby competitors can result from their exposure to chemicals released from the leaf litter and other debris from the source plant as well as exudate from their root system.

Allelopathic chemicals comprise many different chemical groups including carbohydrates, phenolic acids, amino acids and a range of organic compounds that may also serve to ward off herbivores and parasites.

Several basic groups of compounds that serve as the "building blocks" or precursors for much larger molecules are shown in Table 1 (below). While many of the complex compounds formed from these precursors act as biologically interactive natural toxins, several have other critical functions that may or may not serve a protective role. For example, although many carbohydrates perform allelochemical functions they are often primarily considered to be a source of immediate energy for many biota. They may also serve as an energy reserve in the form of starch, or through their role in fatty acid metabolism. Complex polysaccharides (sugars) and other carbohydrates, in the form of cellulose, chitin and collagen, also perform important structural functions related to physical protection and motility. For example, cellulose consists of a string of glucose molecules linked together by hydrogen bonds at the 1,4 positions of adjacent molecules (Table 1). This polysaccharide is the most important structural component of plant cells. It is also produced by a single class of small tube-dwelling aquatic animals known as tunicates. These are members of the phylum Chordata which also includes the vertebrates. As distant relatives of vertebrates, tunicates possess a primitive neural tube, the notochord, but no protective vertebrae. Their strength and rigidity are derived from secreted tubes, or tunics, containing cellulose as a strengthening agent.

In many cases simple compounds such as amino acids can be thought of as building blocks for a multitude of compounds many of which perform the role(s) of natural toxins (Table 1). For example, peptides consist of relatively short chains of amino acids. As their name suggests, polypeptides consist of longer amino acid chains, whereas proteins are essentially polypeptides characterised by a specific (peptide) bond between the carboxyl and amino groups of adjacent amino acids. This sort of cross-bonding among amino acid chains gives proteins and other polypeptides their characteristic shape(s). Such shapes often play an important part in the unique 'lock and key' requirement for accessing specific receptor sites on cells and membranes that influence specialised hormonal or enzymic function. In several cases such molecular configurations may affect their effectiveness as natural toxins. Structural constraints dictate the role of specific amino acids as precursors of alkaloids. In many cases the properties of alkaloids are influenced by specific sidechains and their position on the alkaloid molecule.

Table 1. Examples of Chemical Sub-Groups that are Primary Constituents of Plant-based Compounds with Allelopathic and other forms of Biological Activity.

Phenol. A single hexagonal ring compound with a hydroxyl (OH) group. Polyphenols are a group of approximately 8000 allelochemicals based on the phenol molecule that are found in many plants, particularly nut trees. Examples include flavonoids.	OH Phenol.
Phenolic acid. A phenol-based compound incorporating a carboxyl (COOH) group. They comprise about one third of phenolic allelochemicals. Both polyphenols and phenolic acids have antioxidant properties.	O OH OH Phenolic Acid.
Amino acid. A group of 20 compounds incorporating an amino group (NH_2). 9 are not synthesised by mammals, and must therefore be obtained from food. These are called essential amino acids. Amino acids are the precursors of chained compounds such as peptides, polypeptides and proteins or cyclic compounds such as alkaloids.	H_2N O OH Glycine.

Pyrrolidine alkaloids have a core consisting of two fused 5-carbon rings with a central nitrogen atom which forms the basis for nearly seven hundred Pyrrolizidine Alkaloids and derivative oxides.	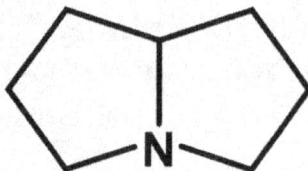 Necine alkaloid base.
Polysaccharides consist of linear chains of as many as 3000 glucose or other monosaccharide units, which give structure and strength to plant, bacterial and some animal tissue.	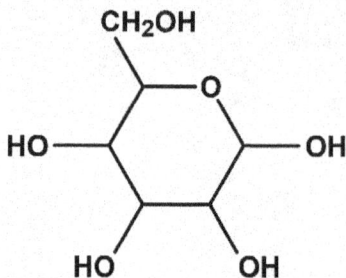 Glucose.
Heliotridine. A pyrrolizidine alkaloid found in leaves and flowers of many plants. It is toxic to a broad range of herbivores.	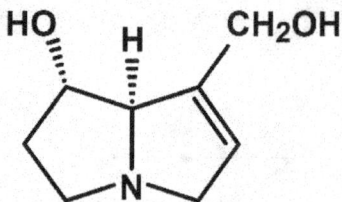 Heliotridine.

Glucosinolates are pungent plant compounds derived from two different constituent groups, amino acids and glucose. that may act as allelochemicals and deterrents to herbivores	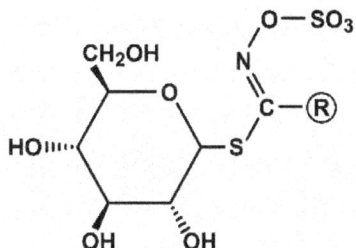 Glucosinolate.
Terpenes and their oxygenated forms known as terpenoids are large classes of aromatic chemical derivatives of the 5-carbon isoprene. Hence their alternative name isoprenoids. Approximately 30,000 terpenes and 20,000 terpenoids are known. Both of these groups are aromatic compounds present in numerous plants.	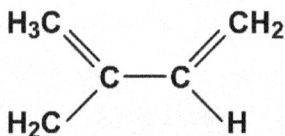 Isoprene.

Two common terpenes are myrcene and β-pinene. **Myrcene and β-pinene** are common natural plant toxin with antibacterial and antifungal properties which acts as a chemical deterrent to insect and other herbivores.

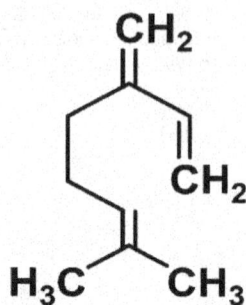

Myrcene.

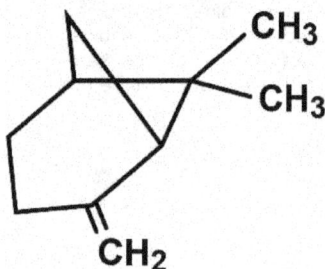

β-Pinene.

An important insight into the relationship between chemical form and function in proteins and other polypeptides was attained through the 2024 Nobel Prize-winning research of Baker, Hassabis and Jumper, who employed Artificial Intelligence (AI) technology to foresee with high probability the 3-dimensional shape of complex molecules based on the make-up and sequencing of component amino acids. This newfound capability has led to an ability to help formulate the degree and nature of molecular folding of a specific compound. This has important implications for the development of novel biologically active chemicals where the characteristic configuration of the compound is critical for its ability to occupy the specific tissue receptor site(s) necessary for its particular biochemical function.

As is the case with chemical defence against consumers, mixtures of allelopathic chemicals may have a greater effect on competing neighbours than an individual compound. Also, their activity may be significantly affected by environmental factors such as solar radiation, temperature, optimal nutrient and water availability. For example The allelopathic potential of creeping vine (*Ipomoea cairica*) has been shown to increase significantly at higher temperatures. Allelopathy is a highly complex phenomenon, as illustrated by the fact that in rare instances allelochemicals may actually have a stimulatory effect on neighbouring plants. Although the effect of these compounds is generally negative, the picture is further complicated by the fact that allelopathic effects are often countered by reactive measures taken by target organisms. On further examination of this phenomenon, it becomes apparent that the notion of 'chemical defence' can be somewhat equivocal, and might be better described as a form of biological warfare wherein terms such as aggressor and defender can be interchangeable. This a theme that will be revisited when dealing with natural toxins and other interactive relationships such as food plants vs. herbivores and predators vs. prey.

An important aspect of allelopathy relates to a dynamic component whereby a plant may alter its allelochemical production according to the degree of prior exposure to different, potentially competitive species. For example, it has been demonstrated that red fescue plants infected by an endophytic fungus produce more allelochemicals than fescue plants free of fungal infection. The time scale of such an increase in allelochemical production is often too short to be considered as a genetic trait established over several generations and may be subject to change over the life span of the plant. However, this does not rule out the likelihood that the biochemical *capability* for *rapid changes* in allelochemical production is itself an inherited genetic trait. In some cases, chemical reaction to an invasive species can be a matter of hours or days. Although we tend to think of plants as 'unthinking', there are many instances of chemotactic behaviour that indicate a significant reactive capacity. Perhaps not a nervous system in the accepted sense but impressive nevertheless!

The word 'competitive' is used in the above (fescue) example even though an endophytic[7] fungal infection does not necessarily result in disease of the host plant. Although the fescue is not damaged by the fungus, it might be assumed that some competition for nutrients might result from such a close physical relationship. However, considering that over 80 per cent of plant species form associations with mycorrhizal fungi, it is not surprising that there are also numerous beneficial effects that can result from such a relationship. For example, metabolic biproducts from plants and fungi in close proximity can be advantageous to both parties as nutrients or as a means of chemical defence against consumers and other common pests such as soil pathogens. In some cases, the presence of fungi can enhance nutrient uptake by the host plant and improve the retention of water from the soil. This property may, therefore, have a part to play in protecting the host plant in the event of drier conditions brought about by climate change. Where mutual benefits can be demonstrated

---

[7] An endophytic plant or fungus grows within the tissues of another (host) plant.

the term symbiosis might be a more appropriate descriptor than competition, and the word 'mutualism' has often been applied to such jointly beneficial relationships.

However, it is important to bear in mind that mutualism can give rise to indirect effects in the event of the introduction of a 'third party' invader. Consider the introduction of a novel invasive species to the neighbourhood of a native plant carrying an endophytic mycorrhizal[8] fungus. If allelochemicals from the new invader are disproportionately more toxic to the fungus, the host plant will lose the nutritive and protective benefits of the fungus, irrespective of allelopathic effects on the host plant itself. Alternatively, there may be no clear advantages or disadvantages that accrue from a close proximal relationship between endophytic species. In other words, the relationship might be described as neutral.

Clearly allelopathy covers a broad spectrum of biological effects in the plant world and may lead to some ambiguity regarding the term 'chemical defence' as it is not always possible to distinguish between an attacker and a defender.

Nevertheless, the term allelopathy is often associated with aggressive invasive plant species, which deploy allelochemicals to supress the growth and reproduction of potential competitors. The word 'deploy' is used advisedly here in the context of plants that are often referred to in the scientific literature as "invasive species". They are characterised by their ability to 'swamp' a new environment, notably at the expense of native plants, especially those that hitherto have had little or no exposure the invasive species. In this context the deployment of allelochemicals is seen as part an aggressive strategy designed to achieve this end. The 'aggressive' analogy extends to one of several theories

---

[8] A mycorrhizal fungus grows within the root system of a host plant.

invoked to explain how these chemical compounds enhance the competitiveness of some plant species over others: namely, the 'Novel Weapons Hypothesis'.

Novel Weapons Hypothesis.

A commonly cited version of this hypothesis states that plant species that encounter a novel invader are particularly vulnerable to the arsenal of allelochemicals carried by that invasive species. A variant of this theory states that intense competition among native species for a limited pool of natural resources will weaken their resistance to an invader that is capable of utilising a different subset of resources available in the same ecosystem. A corollary of this is that allelochemicals are less effective at suppressing well-adapted neighbouring plant species that have shared a similar or close environment with the invasive species over a long period of time. However, some caution should be exercised in rigidly applying this hypothesis unilaterally, i.e. in favour of the new invader. For example, recent studies have reported instances where the 'chemical warfare' favours native plant species. In such cases invaders introduced into a new environment may encounter novel allelochemicals in native plants that *lower* their invasive capacity. In other words there are examples where the Novel Weapons Hypothesis appears to work *against* the invader.

Inconsistencies in theories concerning the history of biological interactions are not new. Even Charles Darwin's Naturalisation Theory contained some seemingly contradictory elements. These include the notion that potential invaders with close taxonomic ties to native species would have a better chance of success than potential invasive species with less co-evolutionary history and less taxonomic commonality than native plants. At first sight this might seem at odds with a New Weapons Hypothesis that favours a completely alien invader.

However there are many conflicting elements capable of shifting the equilibrium among competing organisms. Darwin himself was aware of some of the difficulties in reconciling certain aspects of evolution theory. However, his 1859 landmark publication "The Origin Of Species" indicates a truly remarkable level of insight considering that James Watson and Francis Crick's 1953 discovery of DNA and the genetic code governing the evolutionary process, didn't occur until nearly a hundred years later.

While allelochemicals in the form of natural toxins clearly play an important part in how plants and other organisms influence one another we are still a long way from fully comprehending their function(s). Important insights into their role have been gained from a better understanding of events happening at the ecosystem level. Indeed, it is at this level that much of the research on this subject has been conducted. It is therefore important to appreciate what is happening at the macro level in order to establish the context for events happening at the molecular level.

Following are some important concepts in establishing and defining biological interaction at the macro, or population level.

Invasibility.

"Invasibility" is a term that has increased in prominence, particularly over the last few decades. As the word suggests, it relates to the ability of potentially invasive plants to establish themselves and thrive in a new ecosystem, often at the expense of native plant species. Allelochemicals can clearly play a critical part in the equilibrium reached between the invasive and native species. This phenomenon has been a topic of intense research, particularly over the last thirty years or so, with much of the interest focussed on the deleterious effects of invasive plants on the development of crops, forests and other areas of human interest and activity. In many cases the introduction of invasive species also results from human activity and may have serious economic consequences. In the United States it has been estimated that invasive plants are responsible for a 12 per cent reduction in agricultural crop yields at an annual cost of 34 billion dollars (Pimentel *et al.* 2001).

Non-native species can also result in losses of recreational facilities such as woodlands and waterways (Eiswerth 2005). These in turn lead to reductions in associated goods and services such as forestry products and fisheries. Largely as a result of this intense research interest, particularly over the last thirty years or so, the invasiveness of newly introduced plant species has remained a subject of vigorous discussion and debate. The bulk of this research has followed competing native and invasive plant communities over different spatial and temporal scales. Although the identities of many allelochemicals are well known, most field studies understandably maintain a broad focus on the media responsible for cross-species chemical exposure. The large scale of these experiments doesn't generally allow for precise dose-response data on individual allelochemicals. Typical approaches consist of relatively small-scale experiments measuring the growth and propagation of plants in field plots that are exposed to exudates from potential competitors and soil leachates from locations where competitors have grown for varying lengths of time. Larger scale field studies follow similar lines of enquiry, such as monitoring the invasibility of new plants introduced into zones adjacent to native plants, or into soils with a history of supporting native plant species for varying periods.

Empirical use of exudates, soil leachates and the soils from which they are derived, are not usually accompanied by detailed chemical analysis but may include counts of various microorganisms such as soil bacteria and fungi that might indirectly affect the outcome of invasion by novel plant species. From an economic standpoint, the success of crops and important woodlands in the face of competition from noxious weeds creates a clear incentive for such studies. Consequently, the bulk of this research is performed at the plant community level with a focus on changes in relative dominance among native and non-native plant populations over time. As an illustration of the complexity of the biological interactions relating to invasibility, Enders and Jeschke (2018) cite as many as thirty-three different hypotheses related to invasion biology.

Much of the discussion surrounding this topic relates to the parameters used to define and quantify the success or otherwise of invasive species. Often the endpoints used to measure invasibility are closely related, such as percentage ground cover or biomass achieved by potentially invasive species per unit area over a specified period of time. Other measures of biological interaction may include the relative richness of native and exotic species over time, and from there it is a short step to the numerous published ecological models governing community structure under external stresses. Many of these models are beyond the scope of this book, although it should be borne in mind that allelochemicals and their effects on neighbouring plant communities are at the heart of competition we see at the ecosystem level.

Apart from direct interspecies allelopathy, the success or otherwise of invasives alongside native species may be more closely bound to their relative success in dealing with extraneous stresses such as the harshness of the environment including climate. For example, in harsher climates invasive and native species tend to be taxonomically closer to each other because their ability to deal with environmental change is phylogenetically conserved. Put simply, direct competition between natives and invasives takes second place to the shared genetic make-up that enables them to fight the same war in a difficult ambient environment. On the other hand, direct interspecies competition is likely to be more intense where the surrounding environment is more benign and homogeneous. This amplifies the importance of more subtle species differences that might give either native or invasive species an edge.

Some inconsistences in the determination of invasibility can be reconciled by considering the spatial and temporal scales over which the degree of invasion has been measured. The numbers of native species that are closely related to potentially invasive species seem to be a better predictor of invasion success when counts are made over large scale tracts of land, as opposed to counts made in smaller, more localised field areas where competition for the same limited resources tends to be stronger among closely related species.

Contradictions such as these argue against a narrow adoption of a version of the New Weapons Theory which solely favours the invader and it is clear that the ecological equilibrium related to allelochemical interaction between two species can be shifted by a variety of extraneous influences. For example, a major complication in assessing plant interactions in the natural environment relates to the influence of fungal and microbial communities that may also be part of the same ecosystem and may act as intermediaries between native plant species and invaders. Indirect allelochemical effects on such intermediary organisms may result in a different dynamic between native and invasive plants than would otherwise occur in their absence.

Species Competition and the Role of Natural Toxins.

Following are some important concepts relating to the role played by allelochemicals in the relative success of an invasive plant species in colonising a non-native environment at the expense of native plants.

Allelochemicals are often characterised by the roles that they play in the competition between invasive and native plant communities. The biological interactions of these natural toxins can be both defensive and offensive in nature and have often been described in terms suggestive of biological warfare. In most cases allelopathy relates to antagonistic interactions among plants although, in some reviews, accounts of this behaviour have been extended to the protective properties of these chemical compounds against biocidal microorganisms and other plant pests.

The 'Empty Niche Hypothesis' describes a situation where the increased success of an invasive species relies on its ability to utilize resources in a new environment that are not available to native plants. Unlike the Novel Weapons Hypothesis, invasive success does not result from a direct adaptive response to a native species(s). The invader takes advantage of an ecological niche not utilised by the native plant(s).

The Competitiveness of a novel invader might be enhanced by their ability to better exploit various resources in their new environment than native species. This might involve the deployment of allelochemicals enabling better adaptation to changing environmental conditions and access to more nutrients. This may be seen as outcompeting rather than actively inhibiting native plant species and represents a further illustration of the Empty Niche Hypothesis.

Increased allelochemical production that has been induced because of historical contact with (or proximity to) a previous invader may bolster the inhibitory capacity of a native plant when confronted by a new, non-native potential invader. Allelochemicals are generally less effective against neighbours in the protagonist's own native range. This results from parallel evolution of responses among neighbours to common environmental stresses and probably reflects long-term exposure among native competitors. In such cases, chronic exposure to allelochemicals and consequent chemical countermeasures among competitors would result in significant collective tolerance over time. Such tolerance might be seen as the result of chronic chemical warfare and could be interpreted as the converse of the Novel Weapons Hypothesis.

'Mutualism' is a term used to describe a relationship between native and invasive species whereby both benefit from their proximity to each other. Reasons for the mutual benefit may vary. For example, metabolic end-products from both species may be of joint nutritive value and may supply a greater number and variety of beneficial metabolites than those produced by just one species. Similarly, secondary metabolites from one species might help defend both species against other consumers. In such cases allelochemicals from adjacent species may act additively or synergistically.

Many interactions between invasive and native species may result from indirect effects of allelochemicals from either invasive or native species on other intermediate biota present in the same environment. Examples include the stimulation or inhibition of microorganisms that share the same ecosystem, thereby favouring one or other of the competitors, i.e. either the invasive or native plants.

Defence against herbivores is often regarded as the primary

function of plant-based natural toxins. In many respects this role is analogous to the defence of a prey confronted by a potential consumer, i.e. a predator. Probably no plant herbivore relationship has been studied more than that between plants and insects which share an evolutionary history going back more than 400 million years. However insecticidal plant toxins represent just one aspect of a broad spectrum of competitive biological interactions mediated by natural toxins.

Major Groups of Biologically Active Plant Chemicals.

Terpenes.

Terpenes are a large class of compounds that occur in most natural plant oils. Examples include beta-pinene responsible for the characteristic odour of the pine needles and leaves of related species, and myrcene, which is found in a wide variety of plants, including cannabis, citrus fruit and pomegranates. Myrcene is a common plant terpene found in a wide variety of plants including citrus fruit and cannabis. It acts as a repellent against herbivores such as aphids.

Terpenoids

Terpenoids (see Table 1) are a large group of natural toxins found in both plants as well as fungi and insects. They mostly consist of 20-C structures with a variety of side chains. Their typical core structure is illustrated by the terpenoid kahweol (Figure 3.), characteristic of coffee beans. However, other terpenoids have chemical side chains including alcohols, aldehydes and alkaloids. Over 350 of this family of compounds have been identified in plants. Their pungent or bitter flavour odour often renders plants distasteful to potential consumers, although many also have antimicrobial and antifungal activity.

Diterpenoids are one of several sub-groups of terpenoids found principally in plants and fungi, although they have also been isolated from invertebrates such as insects and marine sponges. They feature in a wide variety of plants such as eucalyptus, cinnamon and cloves that are well known for their aromatic qualities. Many of these herbs have long been sought after for their aroma and taste as well as a broad range of medicinal properties including antibacterial and antiviral activity.

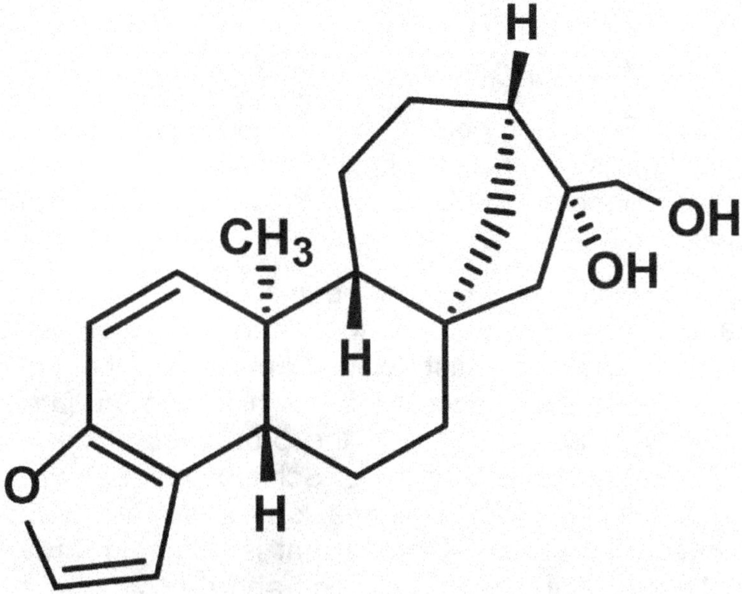

Figure 3. Kahweol. A terpenoid found in coffee beans.

Sage plants (*Salvia* spp.), rich in terpenoids and terpenes, are common in many herbal remedies going back hundreds of years, including instances of psychoactive properties. For example, the Mexican sage *S. divinorum* has a long history as a spiritual hallucinogen associated with shamanism which is reflected in its continued use in some societies. Although the primary psychoactive agent of many *Salvia* species has been identified as the diterpene Salvorin A, it is important to add that other active ingredients include flavonoids and phenolic compounds may contribute to their pharmaceutical and narcotic activity. Their reputation as "cure-alls" for health problems ranging from digestive and heart problems to mental acuity probably relates to the veritable cocktail of natural toxins contained in these members of the mint family.

The Story of Endod.

The African Soap Berry plant *Phylolacca dodecandra* is native to tropical parts of Africa, notably Eritrea and Ethiopia, where the name 'endod' is derived from the Afro-Ethiopian language Amharic.

For me, the story of this plant begins with a visit paid to my Chesapeake Bay laboratory in Maryland, U.S.A. in the 1980s by the Ethiopian scientist Aklilu Lemma. Dr. Lemma had achieved fame and accolades in his own country and elsewhere for his discovery of the properties of the African Soap Berry plant as a remedy for Schistosomiasis, otherwise known as bilharzia. Schistosomiasis is a disease caused by the parasitic worm Schistosoma that is common throughout parts of Africa, Asia, South America and the Caribbean. The pivotal role played by the natural toxin of the Endod plant was that of an effective molluscicide that stopped the spread of schistosoma by killing the freshwater snails that acted as hosts for the worm.

Dr. Lemma had spent an extended period of time in the Chesapeake region on the east coast of the United States. , including a period of postdoctoral study at Johns Hopkins University in Baltimore, Maryland and, as a consequence, he was familiar with my laboratory as part of the University of Maryland system. At the time of Dr. Lemma's Maryland residency, the Chesapeake Bay area was preoccupied by invasive molluscs of its own. The invaders of concern were not disease-carrying gastropods of African origin, but prolific bivalve migrants that originated from the Aral and Caspian Seas between eastern Europe and western Asia. Zebra mussels *Dreissena polymorpha* and Quagga mussels

64

Dreissena bugensis probably arrived in North America from these Eurasian inland seas in the late 1980s. It is believed that these bivalves were transported as waterborne larvae in the ballast water of transatlantic ships or by hitch hiking as adults or juveniles attached to ships' hulls. Zebra mussels were first recorded in Lake St Clare in the north American Great Lakes region in June 1988. Quagga mussels appeared in the Great Lakes shortly thereafter. Zebra mussels were the dominant species, and by 1993 had spread to most of the inland river systems in North America including the Mississippi River and the Susquehanna River which is the primary tributary of the Chesapeake Bay. From a physiological standpoint the likelihood of the spread of zebra mussels throughout the Chesapeake Bay region was aided by the nature of their original habitat. Both the Aral and Caspian Seas are characterised by low and variable salinities, similar to large expanses of the Chesapeake Bay, where salinity ranges from 100% seawater in the south to freshwater in the Susquehanna River estuary in the north. Within this euryhaline[9] environment optimal spawning and survival of zebra mussels occurred at salinities between approximately 10-25% seawater (Setzler-Hamilton et al. 1997).

Concern over the spread of zebra mussels throughout low salinity regions of the Chesapeake Bay region had more to do with potential blockage of municipal and industrial water intakes rather than the direct adverse effects on other biota. In their established European and Asian habitats Zebra mussels were notorious for their clogging of power stations, cooling systems and other coastal industrial facilities. This activity contributed to an annual cost caused by aquatic nuisance

---

[9] Euryhaline refers to (generally coastal, estuarine) areas where the salinity of the water varies from full strength seawater to low salinity, brackish water.

species that added up to tens of billions of dollars. A major defence against invasive aquatic species continues to include their elimination from ballast water and other agents responsible for their introduction into new vulnerable environments.

My interaction with Dr, Lemma focussed on the success of the endod plant product as a molluscicide, and prompted an exploration of endod extract as a means of controlling molluscan nuisance species in water intakes and ballast water (Wright and Magee 1997). However, aside from the battle against invasive nuisance species in the aquatic environment, the prime use of endod continues to be as a natural product molluscicide with the potential for crop protection by its control of herbivorous snails. This application continues to drive its continued cultivation in some countries.

Active ingredients of endod are triterpene glycosides known as saponins. These are secondary metabolites whose structure typically consists of a steroid "backbone" consisting of 4 or 5 carbon rings with chains of sugar molecules attached. The primary endod saponin, lemmatoxin ($C_{48}H_{78}O_{18}$), appropriately named after Aklilu Lemma, is mainly considered as a molluscicide, although antimicrobial and antifungal properties have also been ascribed to this natural toxin. This broad range of properties has contributed to its cultivation and use as a cheap, readily available remedy for the treatment of such diverse afflictions as ringworm, gonorrhea and intestinal worms. These pharmaceutical properties, together with its joint lipid-soluble and water-soluble characteristics and pleasant aroma have also resulted in its use as a soap and laundry product going back many hundreds of years. Hence the reference to endod as the soap berry.

Cannabinoids.

Some natural toxins are difficult to categorize. Examples include cannabinoids, a group of more than 100 compounds that have been isolated and recognised for their psychoactive properties since the late 1800s. However, the therapeutic use of cannabis plants has been documented in Chinese texts as old as 2800 BC and has since been recorded in ancient Hindu, Roman and Greek texts. These documents speak of the use of cannabis as an analgesic and muscle relaxant, and a record of the topical application of cannabis as an anti-inflammatory agent appears among 700 Egyptian herbal remedies recorded about 1550 BCE in what has become known as the Ebers papyrus[10].

Cannabis is probably best known, however, for its mood enhancing qualities including the promotion of a euphoric state and for the relief of anxiety and depression. Some texts have described these effects as "forgetting your troubles". A putative example of such a palliative effect of cannabis appears in Homer's Odyssey written in the early 8th century BC, where Zeus' daughter Helen prescribes a mixture of a narcotic mixed with wine to be given to Greek soldiers mourning the loss of their comrades in the Trojan War. The drug is described as nepenthes, of Greek origin, which translates as "Nē–Penthés" or "No-Grief". This certainly sounds like an invitation to 'put your troubles behind you', and 'Nepenthes' in this case has been widely interpreted as cannabis. However the identity of this drug remains a subject of conjecture and other candidates for nepenthes as described by Homer include opium.

it is easy to see how the medical and recreational uses of

---

[10] The historic Egyptian papyrus of medical knowledge dating from ca. 1550 BC was bought by the German Egyptologist Geog Ebers in 1872. It now bears Ebers' name and is kept in Leipzig University library.

cannabis have become conflated over several centuries and, in several respects, this is still reflected by the equivocal nature of current international legislation governing its use. It is impossible to address the legal status of cannabis world-wide without quickly becoming aware that this is a moving target and rife with contradictions. In the U.K. it is illegal to possess, grow or distribute cannabis, with maximum prison sentences of 14 years for 'distribution' and 5 years for 'possession'. However, the possession of small amounts of cannabis for personal use often results in no more than a warning. In 2020 registration of Cannabidiol on the U.K. medical registry was prompted by the successful treatment of seizures caused by severe cases of epilepsy. Several cannabis-based products continue to be developed for similar applications. For example, Sativex is a mixture of tetrahydrocannabinol (THC) and cannabidiol (CND) that has received approval from the U.K. registry for the treatment of spasms resulting from muscular sclerosis. In the U.S.A. recreational use of cannabis is currently legal in 24 out of 50 States, although the figure for legal use of cannabis under medical recommendation rises to 38 States as a result of evidence of its remediation of some neuromuscular disorders. Despite being described by the Hindu god Shiva as the 'Food Of The Gods', cannabis remains illegal in India. However, in 2016 Australia legalised the growth of cannabis for medicinal and scientific purposes. In Australia and New Zealand low-THC hemp is legal for human consumption. The legal picture of cannabis world-wide remains 'mixed'.

Cannabis plants, also known as hemp or marijuana, originated from western China and central Asian countries such as northern India, Pakistan and Afghanistan. The two most common species are *Cannabis sativa* and *C. Indica*, both from central Asian origin. Their primary active compound is tetrahydrocannabinol (THC) (Figure 4.) which has a 21-carbon, three-ring structure that is characteristic of cannabinoids, although this is subject to variation in this group of compounds depending on the nature of the chemical side-chain(s).

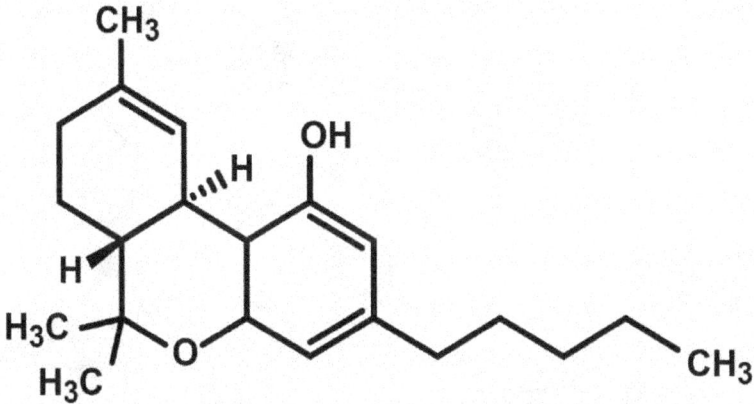

Figure 4, Tetrahydrocannabinol (THC). The primary psychoactive component of cannabis plants.

Cannabinoids have been variously described as multiring phenolic compounds. However, for some, the name terpenophenolic has been employed to capture their hybrid nature with terpene and phenolic components. Interestingly, the degree of psychoactivity in cannabinoids has been shown to be modified by the presence and configuration of terpene components. Although the Cannabis plant is the best-known source of cannabinoids, these compounds have also been isolated from other plants such as Echinacea.

Peptides.

Peptides have an array of protective functions that enhance

the survivability and reproductive performance of plants. For example, they have the ability to confer plant resilience in the face of adverse environmental conditions including drought. Peptides in flowers have been shown to attract bees, butterflies and other pollinators. Several, peptides contribute to plant defence against pathogens. For example, Asperpeptide A (Figure 5.) is an antibacterial, antifungal cyclic peptide found in flowers, leaves, stems and roots of plants.

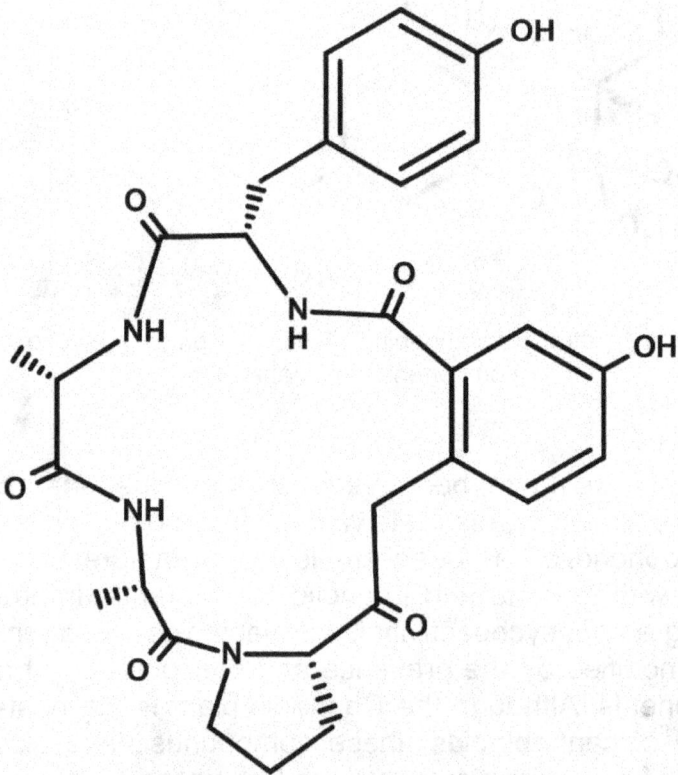

Figure 5. Asperpeptide A. A cyclic antibacterial peptide found in Aspergillus fungi.

Low Molecular Weight Phenolics.

Phenol consists of a benzene ring with a hydroxyl group and forms the basis for numerous biologically active low molecular weight compounds that have been isolated from many plants including citrus fruit and vegetables. These compounds are distinct from the larger molecules that contribute to the chemical structure of cannabinoids already described.

Flavonoids comprise a diverse group of about 6000 phenolic secondary metabolites whose characteristic 3-ring configuration is illustrated by the structure of the common plant compound flavone (Figure 6.). In plants flavonoids have a wide variety of roles, including promotion of seed germination, seedling growth and as olfactory attractants for certain pollinators. Flavonoids also contribute a large range of colours to a broad spectrum of flowering plants. On the other hand, several flavonoids have insecticidal properties. For example, their inhibition of the moulting hormone ecdysone is a highly effective deterrent to potential consumers such as caterpillars. Defensive roles extend to antiviral and antibacterial activity as well as a the Phyto-inhibition of seedlings of nearby competing plants. Some flavonoids have been shown to play a part in the in ability of plants to adapt to changing environmental conditions.

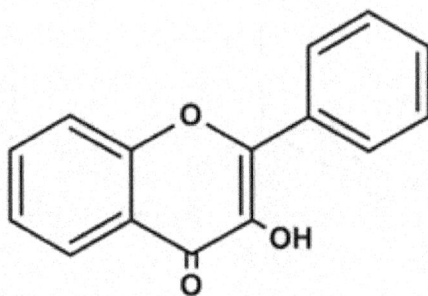

Figure 6. Flavone.

71

Examples include the amelioration of salt stress resulting from climate change. Flavonoids have also been shown to have a protective function against ultraviolet radiation.

Coumarins and Fouranocoumarins

The coumarin group of compounds were first isolated from plants over 100 years ago and their name is derived from the French word for Tonka bean *Coumarouna odonata*, an early source. They have since been found in many different plant families including the Apiaceae (Umbelliferae), the Asteraceae (Compositae) and the Fabaceae (Leguminaceae). The basic bicyclic ring structure of simple coumarins is shown in Figure 7. However, there are many hundreds of variants with different side chains, including the furanocoumarins (Figure 8.), a family of phenolic compounds with a basic structure consisting of a 5-membered aromatic ring (furan) fused to a pair of (6-membered) benzene rings. These compounds enter the nuclei of skin cells where they form cross-linked bonds with DNA under the influence of UV light resulting in inflammation.

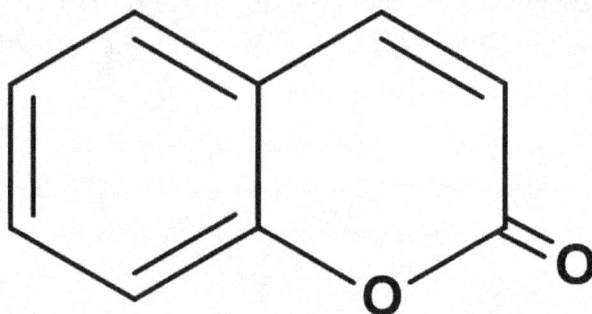

FIGURE 7 Coumarin.

Numerous biological activities have been attributed to simple coumarins and their derivatives and it seems likely that many evolved separately as a defence against predators. In many cases the odour or taste of coumarins and related compounds is sufficient to ward off potential consumers, although several also have antibacterial and antiviral properties. For example, clorobiocin and novobiocin are coumarin antibiotics with an inhibitory effect on the activity of the enzyme DNA gyrase in gram-positive bacteria. Some coumarins have demonstrated anticholinesterase activity, with consequent neurotoxic effects on prospective animal consumers of the coumarin-containing plant tissue.

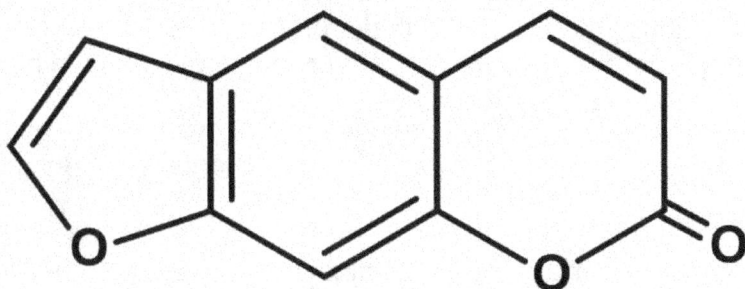

FIGURE 8 Fouranocoumarin

Glucosinolates.

Glucosinolates (see Table 1) are pungent plant compounds derived from both amino acids and glucose that may act as allelochemicals and deterrents to herbivores. They consist of a sulphur-bearing glucose core (thioglucose) linked by the sulphur atom via nitrogen to a sulphate group and a further aliphatic or aromatic sidechain (group R in the structural diagram of Glucosinolates shown in Table 1).

The nature of these sidechains is responsible for the characteristic pungent flavour/odour of the plant which can act as a deterrent to consumers. The production of these chemicals can be exacerbated by damage to plant tissues caused by herbivorous animals. They are characteristically found in plants of the order Brassicales (e.g. cabbage, broccoli, horseradish, radish)

Alkaloids.

Alkaloids comprise an enormous class of naturally occurring chemical products that are derived from a broad range of different organisms, particularly plants. These compounds constitute perhaps the most widespread single class of natural toxins in the plant kingdom.

Natural Function(s) of Alkaloids.

The toxicity of many plant alkaloids is seen as a means of defending against predators, consumers, and competitors. Defensive 'strategies' may include the bitter taste associated with several nitrogen-containing groups characteristic of many alkaloids. This will act as a simple deterrent to grazing animals, although in many cases their deterrent function extends to much more serious biological effects including acute and chronic toxicity to a large range of potential consumers. In this role it has been shown that alkaloids from different parts of a plant, can be transported to specific tissues to fight the depredations of consumers such as insects, grazing livestock and bacterial and fungal infestation. For plant protection alkaloids are usually concentrated in edible leaves and stems, although flowers are not exempt, and the deterrent function of these compounds has been clearly demonstrated against a variety of plants pests and consumers. Several alkaloids found in the Solanaceae (nightshade) family including solanine and nicotine are highly toxic to a variety of chewing insects as well as sucking insects such as aphids and whiteflies. Antimicrobial activity has been demonstrated by several plant-based alkaloids such as piperine and tomatine from black pepper and tomato plants respectively. Berberine from the Indian Barberry (*Berberis aristate*) exhibits both antibacterial and antifungal properties.

The toxicity of alkaloids to potential consumers takes many different forms and the degree and specificity of their biological effects is often dictated by sometimes minor differences in their chemical structure. Such changes in molecular configuration result from the wide range of source chemicals that are now known to be alkaloid precursors. These in turn are subject to a broad range of chemical reactions including acylations, glycosylations, methylations, oxidations, and reductions. These processes have resulted in a massive number of alkaloids, many of them synthetic or modified from natural products, that are responsible for a huge range of biological activities.

The name 'alkaloids' comes from the term "alkaline" which is used to describe any nitrogen-containing base and reflects their common properties as organic bases that form salts with acids and when soluble form alkaline solutions. A common denominator in defining an alkaloid is the presence of a basic nitrogen atom in the molecule, although this is usually identified as part of an integral cyclic structure within the molecule.

Taxonomic Distribution of Alkaloids.

Alkaloids were initially regarded solely as plant products

although have since been found in bacteria, fungi and both invertebrate and vertebrate animals. Many have pharmacological and toxicological properties which have been known for millennia, and alkaloid-producing plants have long been used as pesticides, medicines, poisons and narcotics. As a result of their refinement and adaptation for human use, probably no other class of natural toxins shares such a rich history or has been the focus of so much attention from the pharmaceutical industry. Since their discovery and initial isolation many attempts have been made to categorise alkaloids according to their chemical structure and natural source. However, the sheer numbers and diversity of these compounds has made their classification difficult, and their nomenclature remains a subject of conjecture.

Plants, particularly Angiosperms[11], remain by far the dominant source of alkaloids which have been found in about 25% of all plant species. Plant species belonging to the Amaryllidaceae, Asteraceae, Berberidaeae, Fabaceae (Leguminaceae), Liliaceae, Ranunculaceae, and Solanaceae families have been found to be particularly rich in alkaloids. In the scientific literature the numbers of different alkaloids are usually given as 'over 20,000', although some estimates are as high as 40,000+ different compounds. Despite the apparent disparity between these numbers, they can no doubt be reconciled through the inclusion of countless derivatives achieved naturally or through human manipulation. Much of the history and folklore attached to these compounds is bound up with their use as drugs and medicines, topics that will be dealt with

---

[11] Angiosperms comprise 80% of the plant kingdom. They are seasonal flowering plants with seeds encased in fruits, which are developed from ovaries. Ovaries are the female reproductive system at the base of the flower. Angiosperms differ from Gymnosperms, which are evergreens and do not form flowers. Gymnosperms typically form naked seeds that develop in unisex cones.

in more detail later. However, it is important to remember that many of their biological properties are derived from their roles in the natural environment.

As is the case with many natural toxins the biological activity of many plant alkaloids is seen as a means of defending against predators, consumers, and competitors. Defensive 'strategies' may include the bitter taste associated with several nitrogen-containing groups characteristic of many alkaloids. This may act as a simple deterrent to grazing animals, although in many cases their deterrent function extends to much more serious biological effects including acute and chronic toxicity to a large range of potential consumers. In this role it has been shown that alkaloids from different parts of a plant, can be transported to specific tissues to fight the depredations of consumers such as insects, grazing livestock and bacterial and fungal infestation. For plant protection, alkaloids are usually concentrated in edible leaves and stems, although flowers are not exempt, and the deterrent function of these compounds has been clearly demonstrated against a variety of plants pests and consumers. Antimicrobial activity has been demonstrated by several plant-based alkaloids such as piperine and tomatine from black pepper and tomato plants respectively. Berberine from the Indian Barberry (*Berberis aristate*) exhibits both antibacterial and antifungal properties.

Alkaloids as Narcotics and Medicines.

The toxicity of alkaloids to potential consumers takes many different forms and the degree and specificity of their biological effects is often dictated by minor differences in their chemical structure. Such changes in molecular configuration result from the wide range of source chemicals that are now known to be alkaloid precursors. Several of these are synthetic biproducts that result from biochemical manipulation including the use of specific enzymes and genetic engineering techniques. These processes have resulted in a massive number of alkaloids that are responsible for a huge range of biological activities, many of which are related to their role in the natural environment. Inevitably, many of the properties that have benefitted the organisms producing natural toxins have been refined and adapted for clinical and medical use over many years. The earliest uses, often several hundred years old, stem from empirical herbal remedies and folk medicine handed down from generation to generation before the underlying biology and chemistry was understood. A particularly famous example is the hemlock plant (Figure 9.), whose narcotic properties have been known for millennia, but whose toxicity has been responsible for numerous deaths of consumers, both accidental and deliberate. Many poisons have been shown to have therapeutic benefits in sublethal doses, yet in many cases we are again reminded that the difference between favourable and toxic effects of natural toxins is simply a matter of dose.

FIGURE 9 Hemlock.

Hemlock: A Case Study in Poisonings. From Socrates to Shakespeare and Beyond.

Many natural toxins are found in plants and it is, therefore, not surprising that the majority of deaths and sickness result from their accidental ingestion by herbivores, including humans. A particularly famous example is the poison hemlock (*Conium maculatum*) is a herbaceous plant in the Apiaceae (carrot) family, native to Europe and North Africa, although it is now widespread worldwide. Most deaths were a consequence of consumers mistaking hemlock for harmless vegetables such as turnips, carrot, cow parsley and wild ginseng. However, in one case a child died after using the hollow stem of the plant as a toy whistle. Case studies from North America between 1979-1992 give accounts of more than 60 hemlock-related deaths during this period although, overall, records of such accidental hemlock poisonings are incomplete and almost certainly under-reported.

Aside from accidental poisoning stemming from non-recognition of the toxic nature of the hemlock plant, the deliberate ingestion of hemlock in full knowledge of its lethal and sublethal effects has a much more intriguing and often sinister history. Perhaps the most famous case of "death by hemlock" was the execution of the Greek philosopher Socrates, who was sent to his death in 399BCE for "impiety and corrupting the youth of Athens" by being forced to ingest the highly toxic poison hemlock plant. In Plato's work 'Phaedo' he describes a dialogue between Socrates and his friends. Socrates describes what is happening to his body as the hemlock takes hold, in particular a creeping numbness due to the blockade of both sensory and motor neurones, with death finally resulting from respiratory failure. These physical symptoms described nearly 2,500 years ago provide an important insight into the Mode of Action of this particular poison, which will be discussed later.

William Shakespeare knew all about the properties of hemlock. Macbeth was written in 1653 and begins with the witch's potion including hemlock and references to natural toxins from a wide variety of biota.

Double, double toil and trouble;

Fire burn and caldron bubble.

Fillet of a fenny snake,

In the caldron boil and bake;

Eye of newt and toe of frog,

Wool of bat and tongue of dog,

Adder's fork and blind-worm's sting,

Lizard's leg and owlet's wing,

For a charm of powerful trouble,

Like a hell-broth boil and bubble.

Scale of dragon; tooth of wolf;

Witches' mummy; maw and gulf

Of the ravin'd salt-sea shark;

*Root of **hemlock** digg'd i the dark;*

Poet John Keats was aware of the toxic effects of hemlock as is shown in his **Ode to a Nightingale**:

My heart aches, and the drowsy numbness pains

*My sense, as though of **hemlock** I had drunk*

Clearly Keats' poetic perspective refers to the effects of hemlock in terms of 'drowsy numbness', which certainly echoes Plato's narrative describing the creeping death of Socrates. However, most accounts of sickness and death resulting from hemlock ingestion describe many other symptoms in addition to increasing pervasive numbness. These include uncontrollable shaking, convulsions, blindness, asphyxiation, and respiratory failure. It has been suggested that the more subdued passing of Socrates as described may have resulted from the addition of opium to the lethal concoction that he was forced to take. The combination of hemlock and opium was already known to be a recipe for suicide in Greece at the time of Socrates. Some recent theories on the suicide of the Queen of Egypt, Cleopatra in 30 BCE prefer the ingestion of a fatal mixture of hemlock, opium and aconite over a long and painful death inflicted by an Egyptian cobra. We will get to opium (and snake venoms) later. Two chemical types are primarily responsible for the toxic effects of hemlock. Both are neurotoxins that act in different ways, and serve as models for the way toxins, natural or otherwise, operate. These two compounds are cicutoxin the primary toxin of the water hemlock *Cicuta douglasii*, and coniine which takes its name from the poison hemlock *Conium* and is therefore credited with the death of Socrates. The mode of action of these natural toxins is described below.

Cicutoxin, (neurotoxin of water hemlock).

Cicutoxin is a long chain molecule known as a fatty alcohol with the formula $C_{17}H_{22}O_2$. (Figure 10). The compound was first isolated in 1915 and, while its mode of action is still being elucidated, it has been identified as an antagonist of the neurotransmitter Gamma Amino Butyric Acid (GABA), effectively blocking GABA from its normal (GABA) receptor, which is an ion channel receptor controlling the passage of ions across a neural membrane. In its normal function GABA interacts with the receptor to unlock the "gate" that allows charged atoms or molecules (ions) such as Na^+, K^+, Ca^{++}, Cl^- to travel through channels in the neural membrane. These channel proteins facilitate the passage of ions across the membrane. GABA therefore acts as a key to the ionic channel, thereby controlling the transmembrane electrical potential difference (polarity) responsible for propagating the nervous impulse. This in turn determines the degree and rate of neuronal excitability.

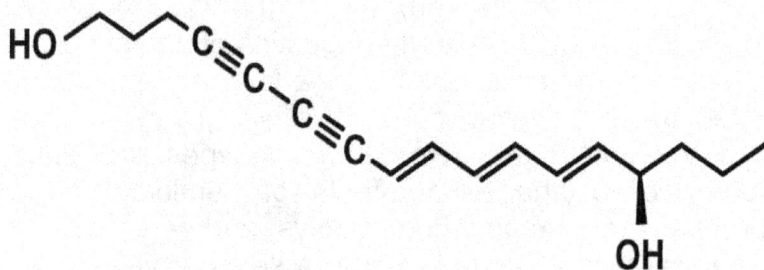

Figure 10. Cicutoxin. An active ingredient of Water Hemlock.

In normal function, following activation by GABA, GABA receptors are responsible for opening the channel causing Cl^- ions to flow across the membrane. If cicutoxin binds to the receptor instead, the receptor is not capable of being activated by GABA, so the pore of the receptor will not open. This is analogous to the 'wrong key being stuck in the lock', thereby blocking the passage of Cl^- across the membrane. This causes constant depolarisation of the nerve membrane which results in uncontrolled hyperactivity and the whole muscle or nerve goes into seizures. Other types of channels can be thought of as revolving doors where the motive force causing the door to revolve is energy-consuming, i.e. active transport. Normally, under passive (i.e. "non-active") conditions, things that will drive chemical compounds or electrolytes across a membrane, will be the relative electrical charge and/or the ionic concentration gradient across the membrane.

For example the passage of sodium ions (Na^+) will be enhanced if they flow "downhill" towards the negatively charged side of a particular membrane. Or Na^+ transmembrane movement will be enhanced simply by a concentration gradient, whereby, all other things being equal, Na^+ just like any other chemical will move across the membrane from the higher concentration to the lower concentration. Things are complicated by the fact that electrical potential difference on either side of the membrane may hold back a chemical against its concentration gradient. This potential difference essentially provides the motive power that drives the transmission of a nerve impulse at a speed of 100M per second! In humans, hundreds of billions of neurons each make thousands of junctions (synapses) with other neurons, i.e. an estimated trillion active synapses. Any blockage of a channel or a transmembrane transport mechanism will destroy the ordered transmission of the nerve impulse.

As a rule, these delicate balances characterising ionic concentration and electrical gradients are maintained by active transport mechanisms requiring metabolic energy, with the added refinements such as ion exchange mechanisms which swap Na^+ and K^+ ions. Under normal conditions the 'outside' (extracellular side) of a cell membrane contains excess Na^+ ions, and the inside (intracellular side) contains excess K^+. Normally, at rest, the outside of the membrane is positively charged, and the inside is negative. This is known as the 'resting potential'. When a stimulus exceeds a threshold level, ion channels open and Na^+ ions are allowed into the cell causing it to become depolarised. Normally this happens in a controlled manner during the passage of a neural impulse but discontinuity of normal electrolytic flow results in the permanent depolarisation of the neural membrane leading to seizures. It is this activity that is triggered by channel blockers, of which cicutoxin is just one example.

Further complications arise from the fact that physiologically useless (or even toxic) chemicals in sufficient quantities can substitute for a physiologically active chemicals in their transmembrane transport. These will be discussed later, with reference to specific chemicals whose biological activity and/or toxicity results from their competition with the substrates that occupy membrane channels and transporters as part of normal physiological function. Examples include the replacement of the sodium ion Na^+ with a unipolar ion such as lithium ion (Li^+) on a carrier molecule Another example is the substitute of calcium ions (Ca^{++}) by cadmium Cd^{++}, thereby inhibiting the role of calcium as a co-factor in several enzyme systems involved in protein metabolism as well as its critical skeletal function in many organisms.

A schematic showing the various ways that electrolytes cross cell membranes, or maybe inhibited from doing so, is shown in Table 2.

Coniine. (neurotoxin of poison hemlock).

The other compound responsible for the toxic properties of the hemlock plant, specifically the poison hemlock, is the highly toxic alkaloid coniine (Figure 11.), based on the single ring piperidine structure incorporating the NH group. Although the alkaloid content in hemlock is often referred to as a single (coniine) entity, the plant actually contains a family of at least

12 similar compounds, all very similar to the coniine alkaloid pictured here. The initial stimulatory effects are similar to the nicotine, but unlike nicotine results in rapid collapse of the nervous system and death of the consumer.

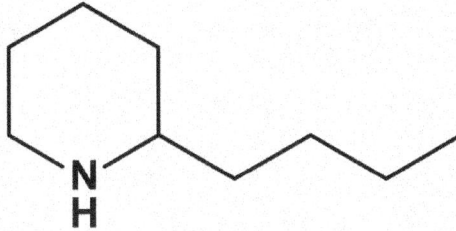

Figure 11. Coniine. A highly toxic piperidine alkaloid found in poison hemlock.

Table 2. Mechanisms Controlling the Passage of Electrolytes and Other Substances across (Neural) Membranes.

Low molecular weight non-polar solutes may diffuse across a lipid membrane or through a membrane pore at a rate dictated by the transmembrane concentration gradient and transmembrane electrical potential difference.

Transmembrane passage of charged ions may occur through a "gated ion channel". Entry to the channel is controlled by Gamma Amino Butyric Acid (GABA). GABA acts as a key to the entry of ions such as chloride (Cl^-), sodium (Na^+) or potassium (K^+) following the attachment of GABA to a specific GABA receptor.

Carrier-mediated ionic transport may be energy driven i.e. 'active transport', and may include ion exchange, where, for example, reciprocal transmembrane movement of sodium and potassium ions occurs via a Na^+/K^+ exchange mechanism.

Pyrrolizidine alkaloids.

Pyrrolizidine alkaloids (e.g. heliotridine. See Table 1) are found in a diverse variety of flowering plants from different geographical regions. Examples include Crotalaria, which is a genus of leguminous plants of the family Fabaceae, primarily from African countries; Senecio (ragwort) a plant genus in Compositae family which is considered to be a native of mountainous/volcanic regions in North America, Africa (Morocco) and Europe; Cynoglossum, also known as the Chinese houndstooth or Chinese Forget-me-not, a genus the Boraginaceae family, considered to be native to Asia and Securinega, a genus in in the family Phyllanthaceae found primarily in Africa but also present in the Mascarene Islands in the Indian Ocean, and in Madagascar where it has achieved notoriety as a highly risky component of a popular coffee substitute made from Securinegra seeds.

The core structure of the pyrrolizidine alkaloids is the necine base, which consists of two five-member rings and a central nitrogen atom (Table 1). This structure forms the basis for numerous compounds several of which are either acutely or chronically toxic to a broad range of herbivorous animals including insects, grazing livestock and microorganisms. In sub-lethal amounts Heliotridine (Table 1) and other plant-based pyrrolizidine alkaloids have been used for the treatment of a wide variety of ailments ranging from nervous and kidney disorders, antiseptic treatment and wound healing.

Bearing in mind that the historic human use of alkaloids and their derivatives has often been more nefarious than controlled medical application, the concept of what constitutes a 'safe dose' can be an important one. Again, we are back to the teachings of Paracelsus. For example, overdosing' can be an unfortunate consequence of recreational drug use. Included here under the rubric 'recreational' is the use of psychedelic drugs as portals to the occult and the spirit world.

Hallucinogenic alkaloids from peyote cactus and psilocybin mushrooms have played important parts in the spiritual rituals of native North American tribes such as the Kiowa and Comanche. The use of hallucinogenic alkaloids is a practice that continues today in some societies throughout the world. In many such communities the distinction between the spiritual and medical use of drugs has become blurred, particularly where genuine curative properties have been demonstrated. Hence the historic use of the title 'medicine man' to describe practitioners in some societies.

Chemotaxonomy of Alkaloids.

Newfound knowledge of the biochemical processes involved with the formation and metabolism of alkaloids continues to inform the structural classifications discussed previously, and enormous recent advances have been made in the biochemistry of these compounds. This massive expansion of our knowledge of alkaloid chemistry has led to a comprehensive reclassification of these compounds based on their biochemical precursors, their biosynthetic pathways, chemical structures, and according to their occurrence in different taxonomic groups of the plant kingdom. These efforts to classify alkaloids have resulted in a seemingly endless array of categories, some with contradictions. This is particularly true in the field of chemotaxonomy which has sought to establish agreement between classical taxonomy and patterns alkaloid distribution in plant groups. However, attempts to establish consistent relationships between taxonomic groups and their alkaloid profile are often ambiguous. Certain taxa are reliable sources of specific alkaloids with similar characteristics. For example, the five major alkaloids including codeine, morphine, noscapine, papaverine and thebaine produced by the opium poppy *Papaver somniferum* L. are grouped under opium alkaloids. However, many alkaloids share a common skeleton within a particular genus of plants but differ in their chemical and biological properties. The occurrence of similar alkaloids among a broad spectrum of plant species makes their classification very difficult, and reviewers are confronted with the question of how to categorise them. Questions include: 'should Natural Toxins be classified according to their chemical structure including common chemical groups, their biosynthetic and metabolic pathways? Or should they categorised according to the taxonomic groups of organisms (mainly plants) that produce them'. Further classifications might include their role, often a protective one, in the natural environment? This topic includes the *raison d'etre* for their evolution, and the genetic selection pressure driving their synthesis and retention by specific groups of organisms. The toxicity of these compounds to a wide variety of biota often reflects their natural protective function in warding off consumers, pests and competitors, and has frequently led to their adaptation and employment as pesticides. They have also been widely used medicines, drugs and poisons.

The pharmaceutical industry dominates much of the published work on natural toxins. In many books and reviews the classification of natural toxins and their derivatives is often dictated by their success as treatments for specific diseases and other pathologies. Often these curative properties are categorised according to the taxonomic groups of organisms that produce these compounds, and several publications focus almost exclusively on the remedial properties of specific biota and the ailments they treat (e.g. 'herbal remedies'). It should be noted that the gene-based selection pressure driving the synthesis and retention of these natural toxins often reflects their protective role in the natural environment by deterring predators, consumers, and competitors of the source organisms.

There is inevitable overlap among these topics leading to frequent cross-referencing between the chemistry and biology of these chemicals, and how their properties have been adapted by humans for a variety of uses. However, a significant feature of this book is an emphasis on _Natural_ Toxins, and their benefit(s) to the organisms that produce them. While their pharmacology in relation to human (and agricultural) medicines and drugs is an important thread in the story of these compounds, it plays only a supporting role in the narrative here. There are several scientific reviews dealing with ongoing development of derivatives of natural toxins for human uses and applications, together with the extensive and complex chemistry and microbiology involved. A few of these publications trace in detail the biochemical and genetic manipulations that have been used to manufacture increasingly effective medicines and drugs. Some are recommended here as reference texts.

While this book explores many chemical groups and a broad range of source organisms, a common theme has been to focus on the interaction between biota and toxic chemicals in the natural environment. In doing so, we consider the effect of natural toxins at different levels of biological organisation from the ecological to the molecular. Reference is made to specific classes of chemicals and biological groups that demonstrate these interactions, but without adhering rigidly to specific chemical groups or biological taxa as overriding themes. As part of this we examine the pitfalls of attempting to classify natural toxins strictly according to their chemical structure, source organisms and pharmacology.

Cocaine. The Good, The Bad and The Ugly.

The distinction between 'good' and 'bad' properties of such compounds can become blurred. For example, the use of the Peruvian Coca plant *Erythroxylum* as an integral part of religious practices goes back more than 1,500 years. Chewing on coca leaves would induce a trance-like state which was presumed to enhance communication with the spirit world. However, by the late 19th century coca extracts gained a widespread reputation as a harmless stimulant with dietary and mood-enhancing benefits.

Endorsements of its use came from several famous people including the scientist Humphrey Davy and the engineer and inventor Thomas Edison. Even in fictional literature the relatively tolerant attitude to cocaine was reflected in its use by Arthur Conan Doyle's famous detective Sherlock Holmes, with the implication that the drug may augment deductive reasoning. Perhaps the most effective 'testimonial' for coca was its original inclusion as an ingredient of the popular drink Coca Cola. Addictive characteristics of the product were downplayed or ignored, although have since been associated with its purified extract, cocaine (Figure 12.). Cocaine is a highly addictive tropane alkaloid, a group that includes nicotine and atropine. It acts as a dopamine agonist by binding to the dopamine transporter. This causes a buildup of dopamine in neural synapses leading to an amplification of the nerve impulse, thereby acting as a stimulant of the central nervous system.

The addictive characteristics of cocaine remain a serious issue in the illegal narcotics trade. There is often a blurred distinction between the spiritual, mind-altering characteristics of some of these compounds on one hand and their often-beneficial qualities as pharmaceuticals on the other hand. It should be added that, despite the echo of the word 'Coke' to describe Coca Cola, cocaine is no longer an ingredient of this, or any other mainstream commercial beverage. However, its place in such drinks has often been taken by another alkaloid, caffeine.

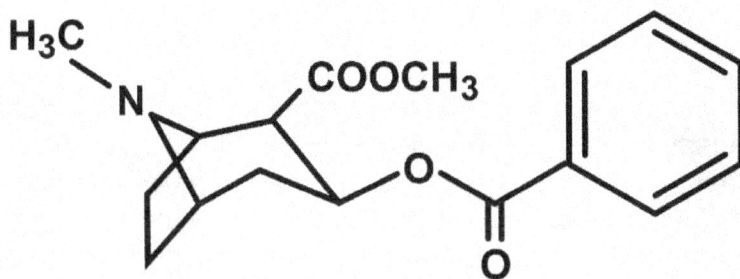

FIGURE 12 Cocaine.

Caffeine

Caffeine has long been a valued stimulant whose use can be traced back to before the Chinese Shang Dynasty (1600-1045 BCE) when infusions of tea plants were a common commodity and used as medications, aids to digestion and for their relaxing properties. However, the initial discovery of the beneficial effects of tea drinking has been attributed to the Chinese Emperor Shen Nung more than a thousand years earlier. In 2737 BCE the Emperor is said to have followed up on the accidental introduction of leaves from the tea plant *Camellia Sinensis* that fell into water that he had boiled as a preliminary to drinking. He recognized the mood enhancing qualities of the concoction and insisted that it became a regular beverage. The Chinese philosopher Confucius is reputed to have been aware of the capacity for tea to concentrate the mind, and from about 500 BCE the medicinal properties of tea its ability to sustain long periods of meditation have been integral parts of 'tea culture' of China and other East Asian countries, notably Japan. Tea drinking has also found an important niche in the meditative process associated with Buddhism. The tea plant itself, *Camellia Sinensis,* is native to the Indian Province of Assam, eastern Burma (Myanmar), northern Thailand, and the Yunnan Province of China.

Although cocaine retains notoriety as an addictive psychoactive drug with spiritual connotations in some societies, it continues to play an important pharmaceutical role as an anesthetic. This dual identity is illustrated by the etymology of the word 'pharmacy', which is derived from the Greek 'Phamakeia', and which has been translated in the King James version of the Bible as 'sorcery'.

However, in looking at the history of cocaine and other alkaloids and their adaptation for human use as medicines, drugs and pesticides, it is important to consider their biological effects in the context of their normal function in the natural environment; the part they play in enhancing the survival and reproductive potential of the organisms that produce them. Their numerous and varied toxic properties are seen as largely protective in nature, acting as a defense against potential predators, consumers and other pests likely to adversely affect their survivability. For example many alkaloids have antibacterial, antifungal and antiviral properties

Since the discovery and initial isolation of alkaloids many attempts have been made to categorise these compounds according to their chemical structure and natural source. However, the sheer numbers and diversity of these compounds has made their classification difficult, and their nomenclature remains a subject of discussion Several differences in definition are still apparent in published literature on this topic. The term 'alkaloid' is normally applied to compounds where the nitrogen atom (N) is part of a cyclic component of the molecule. This definition appears to rule out compounds with N in the exocyclic position, i.e. where the nitrogen atom N is the constituent of a side chain. Likewise, nitrogen-containing compounds such as amines, nucleic acids and nucleotides are generally not referred to as alkaloids. However, the boundary between the term 'alkaloids' and some other nitrogen-containing natural compounds remains blurred. For example, some authors regard some alkaloids as a category of amines. Amines are derivatives of ammonia NH_3, wherein one or more of the hydrogen atoms are substituted by chemical groups such as CH_3 or a benzene ring. In trimethylamine (Figure 13.) the nitrogen atom appears in the centre of a pyramidal structure linked to a trio of CH_3 groups but no benzene ring.

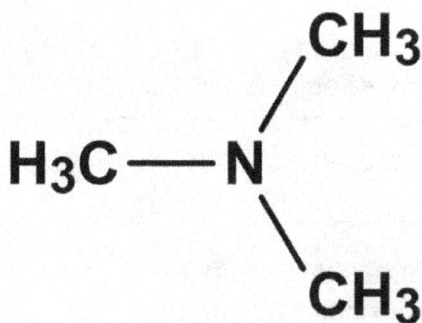

$$H_3C-N \overset{\displaystyle CH_3}{\underset{\displaystyle CH_3}{<}}$$

FIGURE 13 Trimethylamine.

Alkaloid History and Classification.

The name "alkaloids" was first coined more than two hundred years ago by the German chemist Carl Meissner in 1819 who applied it to alkaline plant-based products derived from wood-ash that exhibited a broad range of biological effects on organisms exposed to these compounds. The name 'alkaloids' comes from the term "alkaline"[12] which is used to describe any nitrogen-containing base and reflects their common properties as organic bases that form salts with acids and when soluble form alkaline solutions. As previously stated, a common denominator in defining an alkaloid is the presence of a basic nitrogen atom in the molecule, although this is usually identified as part of an integral cyclic structure within the overall molecular.

A critical aspect of the study of natural toxins such as alkaloids and their biological effects is the history of their discovery, isolation, extraction and biochemical transformation leading to their medical, spiritual and pharmaceutical use. This history

---

[12] pH is logarithmic scale from 0-14 denoting the negative decimal (logarithmic) concentration of hydrogen ions (H^+) in a solution. Historically the term pH describes the *power (p)* or concentration of H^+, hence pH. Although focussing on free hydrogen ion activity the pH scale signifies the relative concentration(s) of H^+ ions and hydroxyl ion (OH^-) concentrations, such that, at pH 7 the concentrations of the hydrogen and hydroxyl ions are equal. Under these conditions the solution is neutral. At pHs less than 7.0 there is an excess of hydrogen ions, signifying acidity. At pHs 7-14 OH^- ions predominate and the solution is said to be 'basic' or 'alkaline'.

also includes more nefarious uses such as murder (state-sanctioned or otherwise) and suicide with all the cultural ramifications these have entailed, including a rich fictional component.

The first half of the 19th century was a highly productive period of discovery and isolation of numerous plant-based compounds which displayed a broad range of pharmacological and narcotic properties. Prior to that time chemicals derived from plants were primarily acidic in nature e.g., oxalic acid, tannic acid.

Early discoveries identified compounds such as morphine, atropine and nicotine that were derived from amino acids and which comprised a nitrogen-containing heterocyclic ring. However, the isolation of compounds where the nitrogen is not part of a heterocyclic ring, and which do not have amino acids as source compounds has complicated the definition of what constitutes an alkaloid and has led to their division into sub-classes based on their chemical origin and structure. For example Eagleson (1994) differentiates among 'True-alkaloids', 'Proto-alkaloids', and 'Pseudo-alkaloids' according to the following definitions:

'True-alkaloids'. Molecules that are derived from amino acids and possess a heterocyclic ring with nitrogen. The main amino acid precursors for alkaloids are lycine, ornithine, phenylalanine, tryptophan, tyrosine and, more rarely, aspartic acid. Based on their biosynthetic precursor and heterocyclic ring system, these compounds have been grouped into various categories which include indolizidine, piperidine, tropane, purine, pyrrolizidine, imidazole, and pyrrolidine alkaloids (Kaur & Aurora 2015). For example, L-ornithine is the amino acid precursor for pyrrolizidine, pyrrolidine and tropane alkaloids whereas lysine is the source for indolizidine and piperidine alkaloids, a group comprising more than 4000 known compounds. Specific examples include atropine (see figure 1), which is a tropane alkaloid, and nicotine (see figure 26), which is a bicyclic alkaloid consisting of a pyridine and a pyrrolidine ring structure.

'Proto-alkaloids', These are the compounds which contain nitrogen atom derived from an amino acid that is not a part of the heterocyclic ring itself, but is located on a side chain. The amino acids L-tryptophan and L-tyrosine are the principal precursors of this class of compounds, which includes the biological amines ephedrine and mescaline. There a several variations of this alkaloid sub-class as well as some apparent contradictions. For example, there are several instances where compounds defined as proto-alkaloids have no heterocyclic ring at all. Aliphatic quaternary bases such as choline, acetylcholine, muscarine and tryptamine are generally excluded from this sub-class although despite being derived from amino acids and contain heterocyclic nitrogen. Such compounds are usually referred to as biological amines while still satisfying the definition of "protoalkaloids". However, when such "protoalkaloids" occur in the same genus or family as "true alkaloids" to which they are biogenetically related (e.g. mescaline in Cactaceae; narceine in Papaveraceae), then it is common to classify them as alkaloids also.

'Pseudo-alkaloids' do not originate directly from amino acids but can be derived through amination or transamination pathways. Non-amino precursors can also produce pseudoalkaloids such as acetate or phenylalanine. The best-known examples include caffeine, capsaicin and ephedrine. Some are derivatives of monoterpenes, sesquiterpenes, diterpenes and steroids.

Despite these variants in definition, the name 'alkaloids' has persisted as a general descriptive term more than two hundred years after its first use and even includes compounds that are neutral or slightly acidic in nature. Recent advances in analytical chemistry have led to the identification of alkaloids in over 20,000 species of plants.

Evermore sophisticated chemical analytical techniques and computer programmes designed to recognise molecular patterns have enabled the identification of pathways responsible for the biosynthesis and metabolism of specific alkaloids. These tools have led to the identification of alkaloid precursors and the specific enzyme systems responsible for the synthesis and metabolism of these compounds. As more alkaloids have been discovered and more is learned of their structure(s) QSAR (Quantitative Structure-Activity Relationships) analysis has provided insights into their biological function(s).

Current Alkaloid Uses.

The massive and ongoing initiative directed towards the isolation and production of new alkaloids has been mainly driven by efforts to develop novel and more effective medicines and pharmaceuticals from an ever-increasing range of natural sources and the synthesis of products developed for the targeted treatment of specific ailments and disorders. While some alkaloids such as β-carboline (Figure 14) have been synthesised *de novo* from their constituent chemical groups most of this effort continues to be derived from the extraction of alkaloids from plants and other natural biological sources. These products are then subjected to various biochemical transformations often under the control of selected enzymes responsible for specific biosynthetic pathways. β-carboline is a heterocyclic indole alkaloid consisting of an indole group attached to a nitrogen-containing pyrrole ring (right). It is found in some plants but was the first alkaloid to be synthesised *de novo* by German chemist Albert Ladenburg in 1886 from the amino acid L-tryptophan. The structure forms the basis for a family of compounds known as the β-carbolines. These include harmine, which was identified as a hallucinogen in the 19th century, but now forms the basis of a medical treatment of Parkinsons disease.

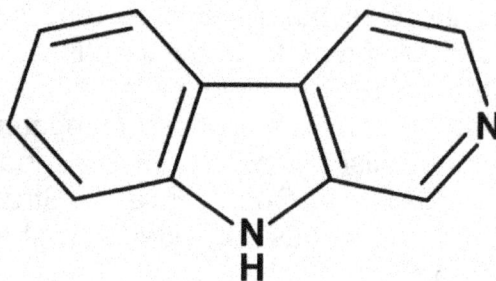

FIGURE 14 β-Carbolene

Through recent advances in molecular technology, biosynthetic pathways may be customised and subject to genetic manipulation to create alkaloids with the desired characteristics and to achieve higher product yields.

The medicinal and pharmaceutical uses of alkaloids is a critically important part of the story of these compounds and has been the subject of numerous books and reviews. Many of these deal at great length with the chemistry of alkaloids and related compounds, their precursors, primary substituents, myriad derivatives and the chemical reactions and substitutions that link these components, particularly as they relate to the pharmaceutical industry. Such a detailed chemistry of alkaloids is largely beyond the scope of this book. Rather, the primary focus here is on the role(s) that alkaloids play in how organisms interact with their natural environment, including both biotic and abiotic spheres. Inevitably human use of these compounds is an important part of this scenario, but it is important to remind ourselves of the obvious, which is that these compounds did not evolve for the betterment of humans and their livestock through beneficial medicines and drugs! Indeed, a large number of animals, humans and their livestock included, are among the very predators driving the evolution of many of the toxic properties that alkaloids possess.

The basic definition of alkaloids has undergone frequent refinements over the last several decades but has remained in use despite some contradictions that have led some to suggest that the term is outdated. The term survives often with some generalisations that may be subject to interpretation.

Nowacki and Nowacka (1965) proposed the following definition:

'An alkaloid is a substance with nitrogen in the molecule, connected to at least two carbon atoms. The molecule must have at least one ring, but it is not necessarily heterocyclic. The compound cannot be a structural unit of macromolecular cellular substances and cannot serve as a vitamin or hormone'.

In addition to information on the chemical structure of the alkaloid molecule an interesting aspect of this definition is the reference to their biological function or rather, in this case, functions that alkaloids do *not* perform i.e. vitamins and hormones. Stated in this way the roles of alkaloids are seen as secondary in nature. They are not regarded as part of 'physical' entities such as proteins, skeletons and epithelia. Neither are they directly engaged in metabolic functions such as digestion, energy generation, locomotion, reproduction etc. In other words, they are not seen as an integral part of the primary biological processes that constitute the normal metabolism of organisms. Nevertheless, as specialised secondary metabolites it is clear that they play an important role in enhancing the survival and reproductive potential of plants as integral parts of their defence system against biotic and abiotic stress.

Newfound knowledge of the biochemical processes involved with the formation and metabolism of alkaloids continues to inform the structural classifications discussed previously, and enormous recent advances have been made in the biochemistry of these compounds. This massive expansion of our knowledge of alkaloid chemistry has led to a comprehensive reclassification of these compounds based on their biochemical precursors, their biosynthetic pathways, chemical structures, and according to their occurrence in different taxonomic groups of the plant kingdom. These efforts to classify alkaloids have resulted in a seemingly endless array of categories, some with contradictions. This is particularly true in the field of chemotaxonomy which has sought to establish agreement between classical taxonomy and patterns of alkaloid distribution in plant groups. However, attempts to establish consistent relationships between taxonomic groups and their alkaloid profile are often ambiguous. Certain taxa are reliable sources of specific alkaloids with similar characteristics. For example the five major alkaloids including codeine, morphine, noscapine, papaverine and thebaine produced by the opium poppy *Papaver somniferum* L. are grouped under opium alkaloids. However, many alkaloids share a common skeleton within a particular genus of plants but differ in their chemical and biological properties. A consistent aetiology linking alkaloid profiles and plant taxonomy is often hampered by the occurrence of certain alkaloids or alkaloid types in unrelated plants. In some cases this may be a consequence of a particular alkaloid nucleus being formed from different precursors found in different plant species. For example the indole-based *Erythrina* alkaloid arises from the amino acid tyrosine, while the nucleus of the indole alkaloid brucine is derived from the amino acid tryptophan. The indole (C_8H_7N) nucleus itself is a weakly alkaline molecule consisting of a pyrrole ring fused to a benzene nucleus (Figure 15.).

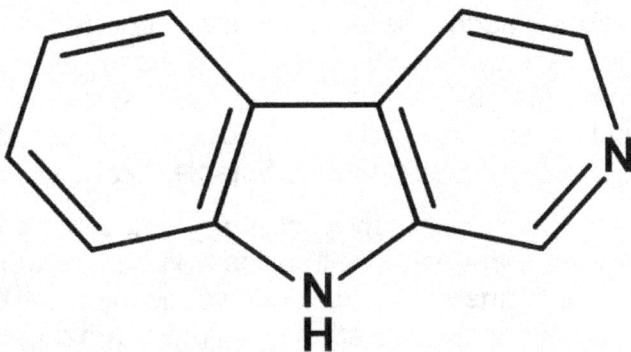

FIGURE 15 Indole nucleus

Pyrrolizidine alkaloids are found in a diverse variety of plants from different geographical regions. Examples include *Crotalaria*, which is a genus of leguminous plants of the family Fabaceae, primarily from African countries; *Senecio* (ragwort) a plant genus in Compositae family which is considered to be a native of mountainous/volcanic regions in North America, Africa (Morocco) and Europe; *Cynoglassum*, also known as the Chinese houndstooth or Chinese Forget-me-not, a genus the Boraginaceae family, considered to be native to Asia and Securinega, a genus in in the family Phyllanthaceae found primarily in Africa but also present in the Mascarene Islands in the Indian Ocean, and in Madagascar where it has achieved notoriety as a highly risky component of a popular coffee substitute made from Securinegra seeds.

The chemotaxonomic picture is further complicated by the sheer numbers of closely related compounds produced from a single alkaloid nucleus. For example, as a result of various biochemical transformations, involving the CYP450[13] and other enzymic pathways, thousands of Indole alkaloids similar in core structure have been identified in several prominent plant families such as Rubiaceae, Loganiaceae and Nyssaceae. In classifications such as these, alkaloids are grouped on the basis of having the same or similar biochemical precursors which, following certain chemical reactions, result in the formation of stable alkaloids.

As part of their defensive strategy against predators, consumers and competitors, alkaloids from different parts of a plant, can be transported to specific tissues to fight the depredations of consumers such as insects, grazing livestock and bacterial and fungal infestation. Defensive strategies may include the bitter taste associated with several nitrogen-containing groups characteristic of many alkaloids. This will act as a simple deterrent to grazing animals, although in many cases their deterrent function extends to much more serious biological effects including acute and chronic toxicity to a large range of potential consumers. An illustration of this activity is seen in the differential distribution of pyrrolizidine alkaloids (PAs) in *Senecio* spp. (ragwort) members of the Asteraceae family. In these plants as much as 80% of the plant's total PA complement is located in the flowers, where the concentration of these alkaloids may be 10-30x higher than the roots and leaves. Experiments by Hartmann and co-workers (Hartmann 1999) demonstrated that PAs synthesised in the *Senecio* roots were transported to flowers in the phloem which is the part of the plant vascular system responsible for transporting the soluble products of photosynthesis. The PAs comprised a group of 360 individual compounds, and it was further

---

[13] Cytochrome P450 enzymes are responsible for detoxifying foreign chemicals (xenobiotics), through an oxidation process that makes them more water soluble and more easily excreted. P450 signifies the spectrophotometric wavelength of this family of compounds.

concluded that the synthetic process responsible for this diverse group was genetically controlled and must have evolved under selection pressure. It might be concluded that the convergence of PAs on the inflorescence (flowers) is part of a selection process that would control the numbers and types of organisms that may be attracted to the inflorescence of a plant but otherwise would have a negative effect on the plant's reproductive capacity by damaging it in some way. Given the positive role that flowers play in the reproductive success of a plant by attracting pollinators, the part played by alkaloids in maintaining an overall reproductive advantage in favour of the alkaloid-producing plant is not straightforward. Complexities such as this, together with the vast array of biological effects displayed by alkaloids, have given rise to a degree of scepticism concerning their overriding protective role. Certainly, plant protection is an easier case to make where alkaloid concentration is focussed on edible leaves and stems (although flowers are not exempt!), and the 'protective' function of these compounds is clearly demonstrated by the antimicrobial activity shown by several plant-based alkaloids such as piperine and tomatine from black pepper and tomato plants respectively. Berberine from the Indian Barberry (*Berberis aristate*) exhibits both antibacterial and antifungal properties.

The toxicity of alkaloids to potential consumers takes many different forms and the degree and specificity of their biological effects is often dictated by sometimes minor differences in their chemical structure. Such changes in molecular configuration result from the wide range of source chemicals that are now known to be alkaloid precursors. These in turn are subject to a broad range of chemical reactions resulting in the massive number of alkaloids that are now known along with their myriad biological activities.

Alkaloids. A History of Human Use (and Abuse).

An early event in the discovery and characterisation of alkaloids was the 1806 isolation of the active "soporific principle" from the from the unripe seed pods of the opium poppy, *Papaver somniferum* by the German chemist Friedrich Wilhelm Sertürner which he named morphine after the Greek god of dreams, Morpheus. Due to its effectiveness as an analgesic and as an anaesthetic, morphine has long been used as an important medicine despite its serious addictive side effects.

The decades following the discovery of morphine saw the isolation of numerous pharmacologically important alkaloids including xanthine (1817), strychnine (1818), atropine (1819), quinine (1819), caffeine (1820), coniine (1826), nicotine (1828), colchicine (1833). The discovery of cocaine was made a little later, in 1860. The pharmacological and narcotic effects of the plant sources of many of these compounds have long predated our knowledge of their chemistry. Indeed, several have been a part of occult and folklore for millennia.

Opium.

The earliest evidence of the human use of opium from seeds of the opium poppy (Figure 16.) dates back more than 7000 years to Neolithic samples from the Mediterranean region, indicating its use in food, as an anaesthetic and, perhaps, as part of ritual practices. A Sumerian reference to 'Hul Gil', the joy plant, which has been attributed to opium, is dated as early as 3400 BCE. Opium is mentioned in many important medical texts of the ancient world, including the Ebers Papyrus (1550

BCE). It also appears in the writings of the Persian physician and philosopher Avicenna in 1025. In addition to occult practices, many of these historic sources describe opium as an analgesic/anaesthetic/sedative for treatment of a variety of ailments/injuries. Historical evidence going back to ancient Greece indicates that opium was consumed in several ways, including inhalation of vapors, and in combination with hemlock for deliberate poisonings, including executions! (see below). Widespread medical use of unprocessed opium continued through the American Civil War, the British Crimean War and the Franco-Prussian War before being superseded by morphine and its successors, which could be injected at a precisely controlled dosage. Opiates may be administered orally, by injection or as suppositories and medical poultices.

FIGURE 16 Opium Poppy

Opioids work by binding to opioid receptors found in the central nervous system and peripheral tissues. Following attachment to a specific receptor they activate one of four different types of transmitters according to the specificity of that receptor. These transmitters are the body's natural opioids:

Endorphins

Endomorphins

Dynorphins

Enkephalins (powerful analgesics)

The activation of these neurotransmitters through mimickry is what affects mood, movement, digestion, pain sensations, and a feeling of well-being. This is what causes the pain relief, euphoria, and relaxation that opioids are known for.

In considering the multifaceted nature of opiates we are, again, confronted with their schizophrenic nature. On one hand we have artists who used opium and its derivatives in the belief that the drug enabled them to relieve ailments and more fully access their creative selves. For examples of those who some may see as exemplifying this this approach we might look no further than Samuel Taylor Coleridge, Charles Dickens and Edgar Allen Poe. On the other hand, we have to consider opiates as an effective group of analgesics with highly addictive properties.

Approximately 12 percent of opium comprises the analgesic alkaloid morphine which is processed chemically to produce heroin and other synthetic opioids for medicinal use and for the illicit drug trade. Heroin was first produced in 1874, and in 1898 The Bayer Company introduced it into the pharmaceutical market as a new pain remedy. Heroin is produced by the acetylation of morphine and is also known by the name diacetylmorphine or diamorphine. Although it was originally marketed on the assumption that it would be less addictive than morphine, it has since become the world's leading illegal narcotic with global use over 20 million. Of the approximately 125,000 annual deaths resulting from opioid use, most are due to heroin overdose. Homicide, caused by deliberate heroin poisoning is relatively rare, but did feature in what, by many accounts, represents the most notorious serial murder case in history. In January 2000, general practitioner Dr. Harold Shipman was convicted of the murder of 15 patients by injecting them with diamorphine (heroin). Although these were the only cases brought to trial, he is estimated to have killed over 250 patients in this way. Dr. Shipman committed suicide by hanging in prison in 2004.

Other opiates derived from the poppy seed pods include codeine and thebaine, which is used to manufacture oxycodone and hydrocodone. Fentanyl, which is often described as a synthetic opiate and is manufactured from a piperidine alkaloid is now responsible for approximately more than 500,000 drug-related deaths world-wide. This is a long way from Paracelsus who, always mindful of dose, prescribed 'tincture of opium' as a sedative and painkiller which he named laudanum.

Monkshood. (Aconitine).

Monkshood (*Aconitum napellus*) (Figure 17.) is another plant species with a long history as both an herbal medicine and a poison. It is also known as aconite and wolf's bane, a name derived from its use as a poison on arrows used to hunt wolves. Common signs of monkshood poisoning include tingling and numbness of the tongue and mouth, nausea and vomiting; breathing becomes harder and laboured; pulse and heartbeat become weak and irregular, and the skin becomes cold and clammy. Death usually results from respiratory and heart failure.

FIGURE 17 Monkshood

The active chemical is the C19-diterpenoid alkaloid Aconitine (Figure 18) which is a potent neurotoxin and cardiotoxin. Aconitine causes persistent depolarization of the neuronal membrane by triggering an uncontrolled flux of sodium ions through channels in the neural membrane. This increases the excitability of the neuron leading to seizures, <u>ventricular arrhythmia</u> and death.

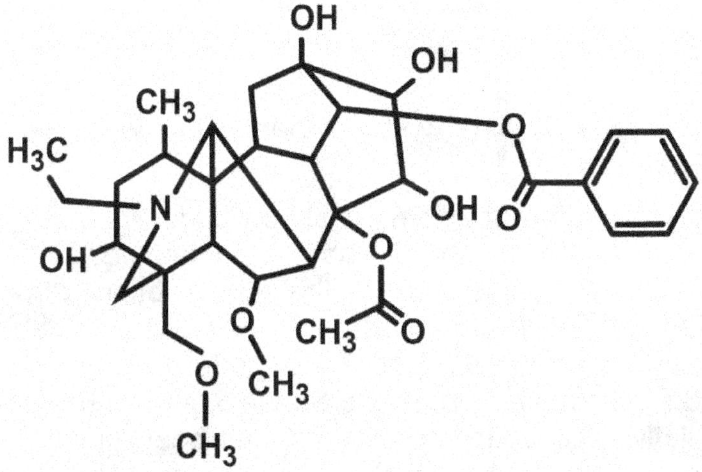

Figure 18. Aconitine. A potent neurotoxin and cardiotoxin causing seizures and death.

Despite its highly poisonous nature monkshood has been used in traditional Chinese medicine for thousands of years for its analgesic, cardiotonic and anti-inflammatory properties. It remains a Chinese herbal remedy for the treatment of several disorders including gastroenteritis, joint pain and rheumatic fever, although the toxic risk remains high and careful preparation is necessary to minimise adverse effects.

Aconitine's reputation as a poison has both historical and mythological components. The Roman god Hercules' encounter with three-headed hound Cerberus, the guardian of hell, is said to have taken place at the hill of Aconitus and resulted in the spillage of saliva, or perhaps vomit, from the mouth of Cerberus. This site was then marked by the germination of a deadly plant which took its name from Aconitus. Another version of the story has the spittle (or vomit) falling on an existing innocuous shrub turning it into the poisonous plant that exists today. Some have imagined that the description of the 'witches brew' from Shakespeare's Macbeth

"Scale of dragon, tooth of wolf, Witches' mummy, maw and gulf ----" (Act 4, Scene 1).

may include a reference to monkshood under its alternative name, wolfsbane. Others have suggested that in Shakespeare's *Romeo and Juliet*, the poison obtained from an apothecary in Verona and used by Romeo to commit suicide in the final act of Shakespeare's tragedy may have been prepared from monkshood. *(Act 5 Scene 3)*. Of course, all this is speculative, although monkshood has been known as a poison since the days of the Roman Empire when it was reputed to have been used as a means of executions and assassinations. The notoriety of monkhood as a poison reached such an extent that its cultivation in ancient Rome was banned.

Cake Anyone?

Perhaps the most famous murder case involving aconitine was the 1881 poisoning by physician Dr. George Lamson of his paraplegic brother-in-law Percy John. In 1882 Dr. Lamson was accused of killing 18-year-old Henry John in order to receive his inheritance. The murder was carried out in Wimbledon, London, during a dinner party, where Lamson is reputed to have demonstrated a novel way of administering medicine using a capsule which he gave to Henry John. As a relatively new invention capsules were employed to allow a noxious-tasting medicine to be ingested without having to taste it. Another version of the poisoning postulated that Lamson incorporated aconitine into Henry John's slice of the Dundee cake Lamson had brought to the party and that was consumed at the dinner table. Part of this theory, however far-fetched, even suggested that aconitine may have been injected into the raisins or sultanas that were part of the cake, or at least the portion that was handed to George Lamson's brother-in-law. Apparently, George Lamson had already left the party by the time Henry John began vomiting and died.

A Matter of Taste.

Whichever device was used to poison Henry John, either capsules or dried fruit (or both?), the object of the exercise was clearly to mask the distinctive flavour of the added aconitine. It is reasonable to ask how we know about the unique taste of aconitine? Before answering that question, it must be remembered that, at the time of the Lamson poisoning case in the 1880s, there were no analytical techniques available that could reliably identify the presence of this alkaloid either in the food or in the body of the deceased. One of the key pieces of evidence in this case was the recorded purchase of aconitine from a pharmacy by Dr. Lamson shortly before the murder. Although this was very good circumstantial evidence it was no substitute for good forensic information physically related to the murder victim and his ingestion of aconitine.

At this point, enter Dr. Thomas Stevenson, an investigator for the prosecution, and a respected authority on the taste of toxins including aconitine. When called as an expert witness Dr. Stevenson was able to testify with a good deal of authority and using his own taste buds. His testimony confirmed the presence of aconitine in the stomach contents and intestines of Henry John's body. It goes without saying that Dr. Stevenson's 'taste test' evidence deserves special commendation. Even if we downplay the sampling of digestive fluids applied to the tongue, we are still talking about the flavour assessment of one of the deadliest poisons ever encountered. Based on this evidence Dr. George Lamson was convicted of the murder of Henry John and hanged at Wandsworth prison in 1882.

The next murder by aconitine poisoning in the U.K. occurred in 2009 when Lakhvir Singh was arrested for poisoning her partner Lakhvinder Cheema when she found out that he intended to marry a younger woman. She was convicted and sentenced to life in prison when it became apparent that she had laced a curry he had prepared for himself and his girlfriend with Indian aconitine. Cheema, known as "Lucky" died, but his girlfriend survived.

Giant Hogweed (*Heracleum antagazzianum*).
Giant hogweed is a very large member of the carrot family (Apiaceae) that can grow up to 4 meters in height. It is native to the Caucasus Mountain region of Eastern Asia, although it was introduced into Europe and North America as an ornamental plant during the nineteenth century. Since then, it has spread to rough pasture and roadside verges where is has proved to be a toxic hazard to both humans and domestic animals. It is often confused with *Heracleum maximum,* the American Cow Parsnip, the only species of *Heracleum* native to North America.

The sap of giant hogweed causes serious skin irritation on contact with all parts of the plant and includes greatly increased photosensitivity which may result in blistering. This condition is known as phytophotodermatitis.   The natural toxins responsible for this reaction are fouranocoumarins (see Figure 8),

Deadly Nightshade, Mandrake and Henbane.

Atropa belladonna (deadly nightshade), *Mandragora officinalis* (mandrake) and *Hyoscyamus niger* (henbane) are members of the Solanaceae family, a large and diverse family of plants which can be traced back to the early Eocene period in the Patagonian region of Argentina 50 million years ago, but is now broadly distributed worldwide. Many members of the Solanaceae are widely collected and cultivated as food plants, including peppers, aubergines, potatoes and tomatoes. However, the inclusion of the three potentially highly toxic plants deadly nightshade, mandrake and henbane in the same family provides a powerful incentive for paying close attention to which part(s) of the plant we choose to consume! The toxic herbs nightshade, mandrake and henbane also provide an illustration of the importance of 'dose', i.e. the amount consumed.

All of these plants contain alkaloids such as scopolamine and hyoscyamine which are potentially highly toxic, but which have anaesthetic and narcotic properties in very low doses. In contrast, the alkaloid atropine (see Figure 1), also commonly found in members of the Solanacaeae can act as a stimulant. Atropine is a tropane alkaloid with medicinal and hallucinogenic properties derived from several members of the nightshade family of plants (Solanaceae). It acts by blocking the effect of the neurotransmitter, acetylcholine.

Deadly nightshade *(Atropa belladonna)*. (Figure 19.). The toxicity of this plant's shiny black berries has been known for several centuries, and the generic name Atropa is derived from the mythological Greek fate Atropos who holds the shears to cut short human life.

FIGURE 19 Deadly nightshade

However, the use of deadly nightshade as a narcotic agent also goes back several hundred years, with applications ranging from sedation to the induction of a trance-like state. As early as the 15th century it was used as a cosmetic, acting to enhance the eye beauty of 'high society' ladies of the Italian Renaissance by dilating the pupils, hence the species name *bella donna* (pretty woman). This property has been retained through the use of atropine in dilating pupils during eye operations.

Mandrake. *(Mandragora).*

The name mandrake has been attributed to several species of the Mediterranean plant genus *Mandragora* including *M. officinarum* and *M. Autumnalis* and has even extended to the white bryony *Bryonia dioica*, otherwise known as the English Mandrake.

FIGURE 20 Mandrake

Since ancient times mandrake has been recognised for its hallucinogenic and analgesic properties and its use as a surgical anaesthetic was initially described by the Greek physician Dioscoredes in AD 60. However, approximately 400 years earlier the philosopher Theophrastus (371-287 BCE) had already written extensively on mandrake as a treatment for gout and sleeplessness, as well as its benefits as a 'love potion'. The consumption of the plant as a narcotic and aphrodisiac goes back to ancient times. In Greek mythology the aphrodisiac properties of mandrake were encapsulated in its description as the "Love Apple" and mentions of mandrake in the book of Genesis in the Bible refer to the power of mandrake to enhance conception, notably that of the sisters Leah and Rachel. When we consider that this Old Testament story is nearly 4000 years old, the history of mandrake is indeed truly ancient.

For many centuries mandrake has been associated with witchcraft and the occult, as well as its proven narcotic properties and its reputed role in promoting fertility. These are all themes that have appeared in numerous works of literature and mythology. Some of the mystique surrounding mandrake relates to the somewhat grotesque resemblance of its root to the human form, although this likeness often requires a stretch of the imagination. The plant's generic name Mandragora is derived from the Latin mandragóras. This, in turn, has deep roots in Greek mythology, reflecting a migration of the teachings of physicians and philosophers from Greece to Rome from about 500 BCE, and recorded in the writings of Pliny the Elder who lived from 23 to 79 AD.

The name mandragora is apparently related to the French 'Main-de-gloire' or 'Hand-of-Glory', which adds a macabre dimension to the story. The Hand of Glory is thought to refer to the hand of a hanged man. Although hands of execution victims were sometimes retrieved as spiritual objects, the

116

mandrake plant, with its bizarre, digital shape was regarded as a substitute for the actual hand of the deceased, and which could be kept as a talisman or amulet. Such a keepsake would ensure the wellbeing of its owner and provide insurance that the recipient of the 'Hand-of-Glory' would not meet the same fate as the donor. However, the possession of mandrake root as a good luck charm clearly did not work for everyone. Joan of Arc is reputed to have carried mandrake to protect her from harm. Nevertheless, this did not prevent her from being burnt at the stake in 1431.

There are several variants of mandrake mythology, including the magical regeneration of new mandrake plants on ground anointed by 'the seed' of the newly executed. However, although it is difficult to conflate the production of semen and death by hanging, we are on more fertile ground when considering mandrake simply as an aphrodisiac. It is no coincidence that the fruit of the mandrake plant are known as 'Love Apples', and that Aphrodite, the Greek Goddess of Love, also has the alternative name Mandragoritis. Nevertheless, these perceived characteristics together with its soporific properties have firmly embedded the mandrake root in folklore as a good luck charm with magic powers to enhance good health and fertility.

The less charitable titles "Devil's Apple" or "Satan's Apple", that have also been applied to mandrake fruit, relate its more dangerous psychoactive properties and reflect a contradictory theme that defines the plant's legacy. The human history of mandrake combines a remarkable range of medicinal properties with an almost unparalleled narrative combining pharmacology and mythology. Both contain good and bad elements. Consistent with the Satanic theme, part of the folklore surrounding mandrake relates to the notional scream of the plant if pulled from the ground; a sound signifying certain death to anyone who hears it. Several mystical characteristics of mandrake have found their way into classical literature.

In Shakespeare's play Henry VI Part II Act III, the Earl of Suffolk, wishing bad things on his enemy invokes the deadly sound of mandrake:

> *Could curses kill, as doth the mandrake's groan,*
>
> *I would invent as bitter searching terms,*
>
> *As curst, as hard and as horrible to hear,*
>
> *Delivered strongly through my fixed teeth,*
>
> *With full as many signs of deadly hate,*
>
> *As lean-faced Envy in her loathsome cave.*

In Shakespeare's Othello (Scene III, Act iii, we find the mischievous Iago sowing seeds of doubt in Othello's mind about the fidelity of his wife Desdemona. Iago states that neither opium (poppy) nor mandrake (mandragora) will give Othello peace of mind in view of Desdemona's suspected adultery.

> *" ----Not poppy, nor mandragora,*
>
> *Nor all the drowsy syrups of the world,*
>
> *Shall ever medicine thee to that sweet sleep*
>
> *Which thou owedst yesterday."*

In his 1630 poem "Go and Catch a Falling Star" the metaphysical poet John Donne makes direct reference to mandrake's fertility-enhancing powers (vis, 'Get with child a mandrake root):

Go and catch a falling star,

Get with child a mandrake root,

Tell me where all past years are,

Or who cleft the devil's foot,

Teach me to hear mermaids singing,

Or to keep off envy's stinging,

And find

What wind

Serves to advance an honest mind.

Although the main theme of the poem apparently relates to his uncertain quest to find a woman who will remain faithful to him, it is possible that the 'cleft in the devil's foot' might be a reference to another physical attribute of the mandrake root. With or without 'the devil's foot' from Donne's poem there remain several literary references ascribing anatomical features to the mandrake plant, including hands, feet and torsos. In addition to these weird physical characteristics of the mandrake root and their associated mystical powers, another enduring magical property attributed to the plant is the deadly scream accompanying its removal from the earth. Thomas Moore's epic poem *Lolla Rookh*, published to great acclaim in 1817, refers to both of these bizarre features of the mandrake plant:

"The phantom shapes – oh touch not them –

That appals the maiden's sight,

Lurk in the fleshy mandrake's stem,

That shrieks when pluckt at night!"

A testament to the longevity of the mandrake legend is its mention in J.K. Rowling's 1998 story of wizardry and magic "Harry Potter and the Chamber of Secrets". In this story the mandrake plant makes an appearance during a herbology class attended by students of Witchcraft and Wizardry at Hogwarts school. Much is made of the need for students to cover their ears in the presence of mandrake, thus avoiding the plant's deadly scream.

Despite the mayhem that has defined mandrake mythology over several millennia, and mixed accounts of its dose-dependent effects on humans, its popularity as a broad spectrum narcotic remained unabated throughout much of the middle ages. Between the 13th and 16th centuries a vigorous trade in mandrake plants throughout Britain and Europe reflected its popularity as an emetic, sedative, hallucinogen and much else, not forgetting its reputation as an aphrodisiac.

The narcotic properties of mandrake are due to a concoction of potent alkaloids including hyoscyamine, l-hyosine (scopolamine) and atropine, all of which are hallucinogenic and soporific in low concentrations but may cause seizures and paralysis in higher doses. An important toxic ingredient in mandrake is the glycoalkaloid solanine which is characteristic of many plants in the Solanaceae family. Like other alkaloids found in the nightshade family, solanine causes a variety of gastrointestinal and neurological disorders including paralysis and death in higher doses. A heavy price to pay for overindulgence!

Black Henbane.

Black henbane (*Hyoscyamus niger*) has been used as a medicine over several centuries and has been described in many traditional medicines. As such it has often been described and employed as a herbal medicine but may induce intoxication accidentally or intentionally. All parts of the black henbane plant, including leaves, seeds and roots, contain alkaloids such as Hyoscyamine, Atropine, Tropane and l-hyosine also known as Scopolamine (Figure 21). Symptoms of acute black henbane poisoning are very varied, and may include pupil dilation, heart irregularity, convulsion, hallucinations and coma.

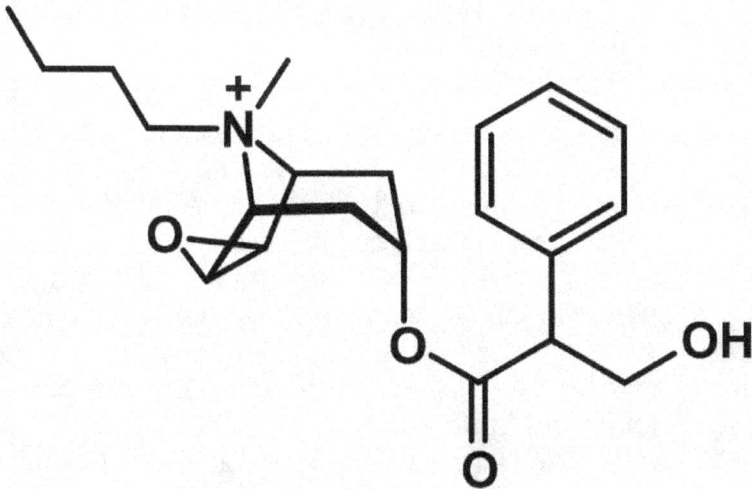

FIGURE 21 l-hyosine (Scopolamine)

Hyoscyamus nigra, the black henbane, is one of the drugs of the ancients and, along with deadly nightshade and mandrake was widely adopted by witches, wizards and soothsayers as a component of their hallucinatory concoctions. In ancient Greece henbane was believed to give people supernatural prophetic powers. It was also used both as a narcotic and poison. It was used by notorious poisoners such as Madame La Voisin in France. Catherine Monvoisin known as "La

Voisin" (1640 – 1680), was a French professional soothsayer and sorcerer, who was implicated in the poisoning of over 2000 people using henbane. Many of these murders were commissioned by members of the aristocracy who sought to gain from the elimination of the victims.

Eventually, in the nineteenth century, the active ingredient of henbane was identified as l-hyoscine, an alkaloid very similar to atropine, which proved to be very important in the study of the autonomic nervous system, and the role of the major neurotransmitter acetylcholine.

The Crippen murder case in 1910 gave hyoscine further notoriety. The unassuming homeopathic doctor Hawley Harvey Crippen allegedly murdered his wife with the alkaloid and then set off for Canada with his young mistress only to be arrested thanks to a telegraph call to the ship as it approached the North American coast. The case was also famous for the forensic chemical and histological evidence leading to Crippen's conviction, i.e. the specific toxic ingredient of henbane, hyoscine (also known as scopolamine) was reportedly found in high concentration in the torso of a body found buried in Crippen's basement. Hyoscine is hardly ever used in modern therapeutics but its history from antiquity to the witches and on to Dr Crippen is both bizarre and fascinating and is still debated. Even as recently as 2007 DNA evidence from the remains of the victim in the Crippen case was still being questioned, nearly 100 years after his execution. Some have puzzled over the wisdom of removing the head and limbs of the victim when the rest of the torso remained buried in his basement!

Datura (*Datura stramonium*). Thornapple.

The Datura plant, otherwise known as the Thornapple, is a highly toxic member of the Solanaceae family whose seeds contain a heady cocktail of atropine, hyoscyamine and scopolamine, all natural toxins with anticholinergic properties capable of blocking the action of neurotransmitters. Although the spiky fruit of this plant have a "Do Not Touch" look and feel, the seeds have long been used for their narcotic properties, but with a very narrow margin separating psychoactive benefits from deadly toxicity and death. Historically Datura seeds have been eaten or smoked as an intoxicant, hallucinogen and occasional aphrodisiac. Indeed, in some societies its soporific tendencies seem to have been favoured by prostitutes as an appetizer for their clients on the grounds that they wouldn't have to work so hard (or at all)!

Differing accounts of the geographical origin of the plant may relate to the fact that there are at least nine different species of Datura spread throughout Asia, Europe and America, sometimes leading to confusion over their common names. Several texts place the origin of the plant in South America, primarily Columbia. However, others maintain that some Datura species are native to Asia going back to the 4th century, with later migration to Arabic countries between the 9th-12th centuries and to Europe as late as the 15th century. Some disagreements remain unresolved and, in some instances, may result from differences in linguistic interpretation.

There is general agreement, however, regarding the potent hallucinogenic effects of Datura seeds. Given the characteristic appearance of the plant and its fruit, accidental ingestion of Datura is very rare. Yet, it was a historic and unplanned consumption of Datura associated with the birth of the United States that gave the plant yet another pseudonym; Jimsonweed.

Jimsonweed.

Since the early venture of the Italian sailor Giovanni Cabot into the Chesapeake Bay region of North America as early as 1497, under the commission of English king Henry VII, this part of 'The New World' became a magnet for European countries with trading and colonial aspirations. Initial forays into coastal Chesapeake Bay areas of what are now the States of Maryland and Virginia were initially limited to relatively short visits and exchanges with indigenous tribes with tobacco as a tradable commodity and the intake of nicotine through tobacco chewing or smoking became a useful catalyst for the 'meeting of minds'. The notion of smoking the 'Pipe of Peace' was more than an invention of Western Movies.

However, in the late 16th century a more aggressive initiative to colonise 'The New World' was mounted by Queen Elizabeth I with Sir Walter Raleigh as her chief military commander. Much of this land-grab had a competitive edge, with European countries such as England, France, Holland and Spain all prepared to compete for territorial and military advantage. During this campaign the wellbeing of native people took second place to the seizure of their land for the growth of tobacco and the theft of livestock and other resources. Raliegh is perhaps best known for introducing tobacco to England, thereby giving nicotine a major boost as a drug for the masses. However, he is also noted for falling out of favour with the monarchy, leading to his 1618 execution by James I for treason.

Back in the colonies, however, things often did not go well. In 1622 the state of Virginia, which had become a major pillar in the tobacco industry, was the site of the massacre of 350 settlers at the hands of the Powhatan native tribe who were convinced that they were fighting for their survival. However, the colonies of European settlers in Maryland and Virginia continued to expand. The English explorer Captain John

Smith had begun exploring the region in 1612 and by 1631 a permanent English settlement was established at Point Lookout in Maryland at the mouth of the Potomac River. Nevertheless, settlers continued to pay a price for their alienation of the native people.

In 1676, following a series of violent clashes with native tribes, the Governor of Virginia, Sir William Berkley was confronted by an angry brandy-fuelled crowd of settlers led by Nathaniel Bacon. Bacon and his rebels had decided to "fight fire with fire"; at least that was their interpretation of the situation. They attempted to solicit Berkley's military support in the planned massacre of the native inhabitants of a Susquehannock village which had nothing to do with any rebellion. Bacon also sought Berkley's assistance in attacking other tribal settlements in the region. Berkley refused and, in response, Bacon's angry mob of settlers reassembled a few months later, marched on the Virginian Capital, Jamestown, and burnt it to the ground. If this sounds like the recipe for more disaster it seems that nature took a hand in derailing further mayhem.

Part of the military force that Berkeley assembled to crush Bacon's rebels never reached their target. In order to sustain themselves before the fight, Berkley's troops adhered to the old adage that an army marches on its stomach and decided to gorge themselves on the seeds of the thornapple plants they encountered *en route*. The seeds of the thornapple plant otherwise known as Datura (*Datura stramonium*) are not very pleasant but there was little else to eat, and the assumption was that this was just another local vegetable. The consumption of Datura turned Berkley's fighting force into a disorganised and disoriented rabble with military discipline dissolving into bizarre hallucinogenic and often childlike behaviour involving conversations with imaginary beings, loss of clothing and playing games with excrement. Normality apparently returned several days later with Berkley's afflicted troops have little or no recollection of their psychoactive detour. The episode earned the Datura plant the name of Jimsonweed, derived from Jamestown, the town at the centre of what became known as the 'Bacon Rebellion'. Again, we find that the psychoactive components of Datura are primarily atropine, scopolamine and hyoscyamine. Irrespective of the intervention of Jimsonweed, the rebellion fizzled anyway following Bacon's sudden death from dysentery in October 1676.

Although accidental ingestion of Datura may result in delirium and hallucinogenic effects, these same symptoms are often deliberately sought after as part of ritualistic practices that continue in some societies. This has resulted in the historic use of Datura and its continued cultivation in some countries. In India, for example, where the word Datura is of Hindi origin, plays an important part in Ayurvedic spirituals worshipping the god Shiva. In some societies Datura is reputed to have aphrodisiac properties.

Mescaline (3,4,5-trimethoxyphenethylamine).

Another hallucinogen with a long and convoluted past is Mescaline, which is a psychoactive protoalkaloid found in the Peyote Cactus *Lophophora williamsii*. This species of cactus belongs to the Cactoideae, which is a subfamily of the Cactaceae family, native to Mexico and southwest Texas.

Mescaline has a long history as both a healing medicine and as a spiritual aid, in which its psychoactive properties have played an integral part in numerous religious ceremonies. Several of these have incorporated aspects of Christianity. However, Peyote has a history as psychedelic and medicine going back more than 5000 years in south American and Mexican culture, and there is evidence of its use in religious ceremonies by indigenous North American tribes long before the advent of Christianity.

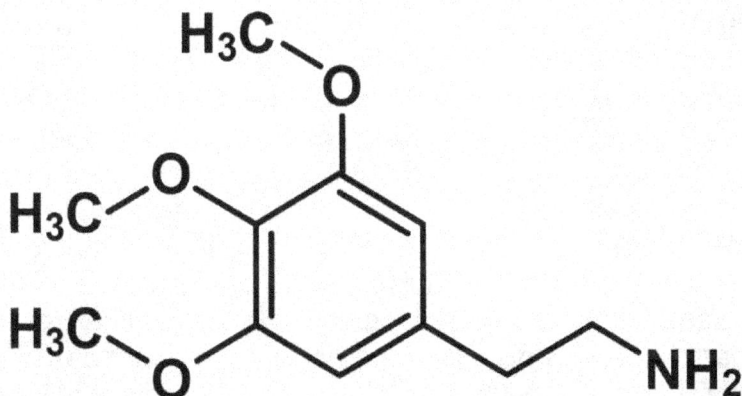

Figure 22. Mescaline. (2-(3,4,5-trimethoxyphenyl ethanamine.

A common denominator in the involvement of Peyote in various religious rituals seems to be its function as a conduit to the spiritual world; a means of communicating with God or gods. This role encompasses a broad spectrum of religious doctrines including Christianity. In this capacity mescaline derived from either chewing or smoking Peyote came to occupy an important position as a revered sacred medicine. Its elevated position as a spiritual sacrament survives in some native North American communities to this day and has resulted in 'Peyotism' being given special legal status among native tribes that have long adopted Peyote as part of their religious belief system. Over 40 North American native communities include Peyote as part of their ritual religious practices. These tribes have included the Navajo, Muscalero Apache, Osage, Kiowa and Lakota. The legal use of Peyote was originally codified as a statute under the American Indian Religious Freedom Act of 1978, and Peyotism now has legal status under dispensation accorded to the Native American Church which was founded in Oklahoma in 1984. Following the passage of amendments to the American Indian Religious Freedom Act of 1994 the cultivation, harvest, possession and consumption of Peyote is legal in the U.S. but subject to certain conditions, i.e. the legal status of Peyote is limited to its use only as part of bona fide religious ceremonies. In a few U.S. states such as Idaho, Utah and Texas the spiritual use of Peyote is restricted to individuals with proven ethnic heritage linking them to specific North American tribes. However, in most cases non-natives are legally able to practice Peyotism as part of proven religious activity or spiritual intent. Otherwise, mescaline, which is the active ingredient of Peyote, is listed as a Schedule I[14] controlled substance under federal law in the U.S. and in Mexico, and as a Schedule III controlled substance in Canada[15].

An examination of how different countries view the classification and legality of mescaline, and its Peyote source makes particularly interesting, if confusing, reading. For example, in many countries, including Denmark, Norway, Ireland, Russia and India the cultivation, possession, transport and sale of psychoactive cacti including peyote are illegal. However, in Slovenia, Sweden, Germany, France, U.K. and Thailand all these activities are legal, although extraction of mescaline from the cacti are either illegal or, in the case of Germany and France, strictly regulated or licensed. In Ukraine, peyote is excluded from banned illicit drugs and is,

---

[14] Schedule 1 drugs have a high potential for abuse and no currently accepted medical use in the country responsible for the classification.

[15] Schedule 3 drugs have less potential for abuse than schedule 1 or 2 drugs.

therefore legal. In countries such as Netherlands and Portugal psychoactive cacti are also legal. In Mexico Peyote is illegal although some other hallucinogenic cactus species such as the San Pedro Cactus remain legal.

In view of the contradictions associated with worldwide regulation of Peyote it is not surprising that some scepticism has accompanied the rules governing its use. An example is the narrow justification of Peyote cultivation "for ornamental purposes" in several countries such as New Zealand. However, a more telling, and somewhat ironic justification for restricting the use of the Peyote cactus had been the recently recognised need for preserving it as an endangered species! Somewhat paradoxically, there is some truth to the dwindling numbers of the cactus *Lophophora williamsii*. However, the primary reasons for this decline have less to do with its use as a sacred medicine, and more due to its recreational use as a hallucinogen. For devotees of Peyote as a religious aid this is seen as unwanted competition, or overharvesting, thereby compromising access to an important spiritual resource. In truth, it is likely that the non-religious use of Peyote has evolved for millennia alongside its spiritual counterpart.

Peyote was first identified as *Echinocactus williamsii* by the French botanist Charles Antoine Lemaire in 1845 but was re-classified in the genus *Lophophora* following the isolation of mescalin 1896. The 1890s saw its transition as a hallucinogen from native North American and Mexican tribes to other American and European cultures, where its ability to induce an altered state of consciousness gave it a broad appeal to artistic communities. This became particularly evident beginning in the 1950s with Aldous Huxley's 1954 work *"The Doors of Perception"*. During the 1950s – 1970s mescalin became a source of inspiration for several authors including William Burroughs, Ken Kesey, Allen Ginsberg and Hunter S. Thompson. In several cases the use of mescalin proved to be an introductory forerunner for the more inexpensive hallucinogen LSD. In addition to its psychedelic effects some medical properties have been attributed to mescalin, including treatment for fever and joint pain (rheumatism). However, solid medical evidence for such remedial powers remains uncertain.

However, in low doses several natural products have been shown to have genuine medicinal properties:

Medicinal use of Natural Toxins.

In low doses several natural products have been shown to have medicinal properties:

Wild Lettuce (Lactuca virosa).

Lactuca virosa is a plant in the *Lactuca* (lettuce) genus, often ingested for its mild analgesic effects and has long been known in the world of "natural remedies" or "alternative treatments". It is related to common lettuce (*L. sativa*), and also goes by the names wild lettuce, bitter lettuce, opium lettuce, poisonous lettuce or tall lettuce. Its use as a drug can be traced back to ancient Greece. It has been referred to as the poor man's opium and was sometimes used by physicians as a sedative in the 19th century when opium could not be obtained. It was studied extensively by the Council of the Pharmaceutical Society of Great Britain in 1911. They discovered two chemicals responsible for the properties of *Lactuca virosa;* lactucin and lactucopicrin. In the United States, the plant experienced a resurgence in popularity in the 1970s. Today the plant is permitted by the Food and Drug Administration (FDA) to be grown and purchased without prescription or license.

Lactucin.

Lactucin (Figure 23.) is a sesquiterpenoid (i.e. C15) compound found in wild lettuce. It has analgesic and sedative properties resulting from its activity as an agonist of the adenosine receptor thereby imitating the role of adenosine as a sleep regulator and arousal suppressor. The plant has long been known to induce a euphoric state when ingested. Although not as strong as opium in this respect it is non-addictive, which accounts for its long history and its role as an alternative to

131

opium. It also has anti-malarial properties. Lactucin has recently demonstrated cytotoxicity in cancer cells, notably lung cancer cells, and its potential for anti-cancer therapy is currently under investigation.

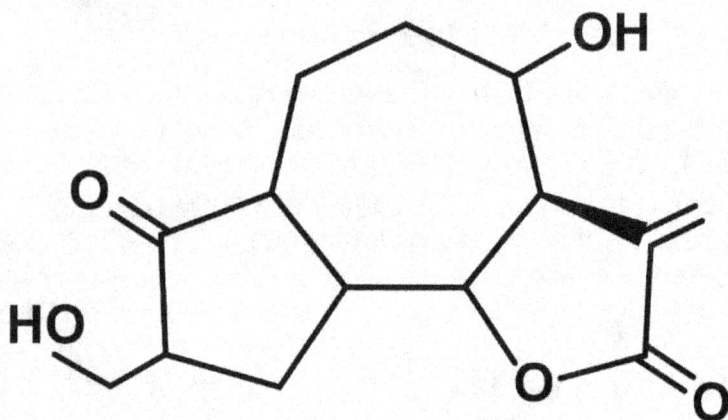

Figure 23. Lactucin. A naturally occurring plant product found in Wild Lettuce (Lactuca sativa) and Chicory (Cichorium intybus L.).

Lactucopicrin

Lactucopicrin (Figure 24) is a psychoactive sesquiterpenoid compound that, like lactucin, has analgesic and sedative effects as well as antimalarial properties. Some applications have reported increases in mental acuity in elderly people including those with to Alzheimer's syndrome. It is found in wild lettuce and dandelion plants. At the molecular level it functions as an anticholinesterase resulting in a build-up of the neurotransmitter acetyl choline at neural synapses resulting in an inhibitory effect on neuromuscular function.

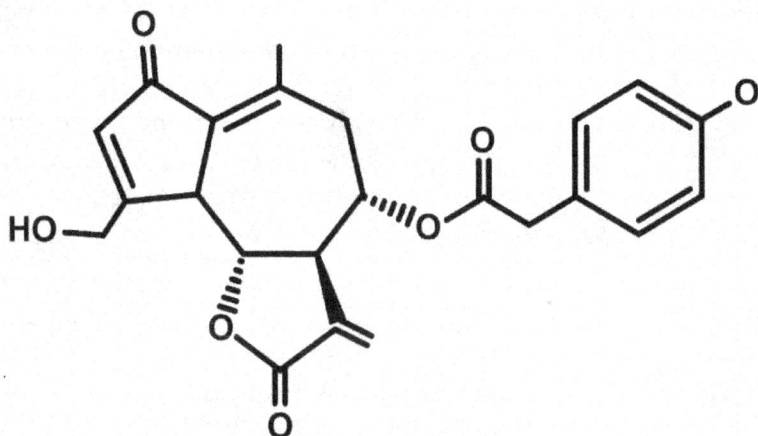

FIGURE 24 Lactucopicrin

Chincona Bark (Quinine).

The bark of the Peruvian Chincona tree (*Chincona officinalis*) is the source of the alkaloid quinine. It was originally native to S. America but is now cultivated in India and Java. Also called Fever Tree, it was used in British colonial India as a treatment for malaria and as a stimulant, considered by many to increase the appetite by promoting the release of digestive juices. It also became known as a remedy for bloating, fullness, and other stomach problems. These qualities were probably the main reason for its adoption as an essential ingredient of a gin and tonic; all for important medicinal reasons of course! However, quinine is also used as a treatment for blood vessel disorders including haemorrhoids, varicose veins, and leg cramps.

Salix Bark (Aspirin).

A precursor to aspirin, found in the bark of the willow tree (genus Salix), has been used for its health effects for at least 2,400 years. In 1853, chemist Charles Gerhardt treated the medicine sodium salicylate with acetyl chloride to produce acetylsalicylic acid for the first time. Over the next 50 years, other chemists established the chemical structure and devised more efficient production methods.

Aspirin, also known as acetylsalicylic acid (ASA), is a nonsteroidal anti-inflammatory drug (NSAID) used to reduce pain, fever, and/or inflammation, and as a blood thinner. Inflammatory conditions which aspirin is used to treat include pericarditis, and rheumatic fever. It is also used long-term to help prevent further heart attacks, ischaemic strokes, and blood clots in people at high risk. Aspirin works similarly to other NSAIDs but also suppresses the normal functioning of platelets.

Experienced practitioners in the art of basketry attest to the fact that, when it comes to weaving willow rods and canes, it's really useful if you have three (or more) hands. Failing that it's not uncommon for strands of cane to be held between the teeth during the weaving process. Basketeers have often noted that a side effect of this activity is some numbness of the lips. What we are seeing here is nothing more than the effects of salicylic acid (aspirin) in its role as a highly effective analgesic.

Side effects of aspirin in high dosages include stomach ulcers, stomach bleeding, and worsening asthma. Bleeding risk is greater among those who are older, drink alcohol. Aspirin is not recommended in the last part of pregnancy and is also not generally recommended in children.

Foxglove.

Extracts of the foxglove plant *Digitalis purpurea* (Figure 25.) have been used for medicinal purposes for hundreds of years, and a written reference to foxglove appears in the herbal register Herbarium Apulueii Platonica as early as the 12th century.

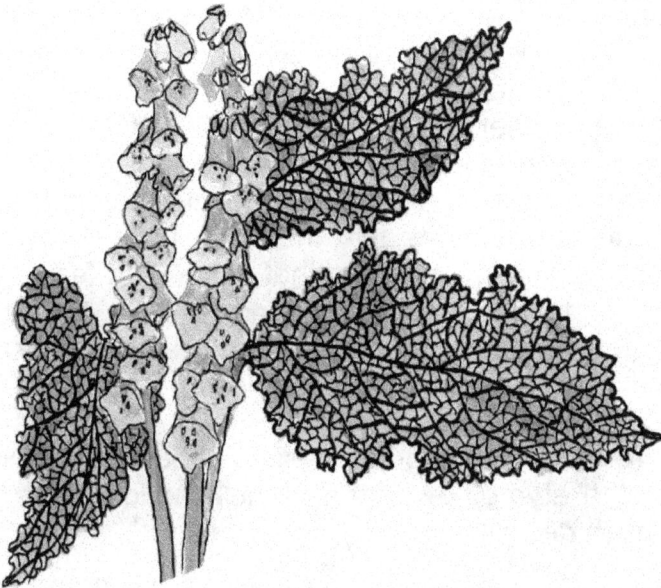

FIGURE 25 Foxglove..

The early use of foxglove extract seems to be primarily as an analgesic, although it acquired a reputation as a cure for oedema and other ailments such as scrophula, a form of tuberculosis. The active ingredient of foxglove was first given the name digitalis by the botanist Leonhardt Fuchs in 1542, although the biggest boost to its medicinal reputation resulted from the 1785 publication of a monograph by William Withering entitled "An account of the Foxglove and some of its Medicinal Uses: with Practical Remarks on Dropsy and Other Diseases". This carefully researched work described the effectiveness of Digitalis for the treatment of heart conditions and is widely regarded as a seminal publication marking the beginning of modern therapeutics. As such, Withering's work is seen by many as a quantum advance from pseudoscientific ideas such as the 'Doctrine of Signatures', which had pervaded herbology during the 15th and 16th centuries. The Doctrine of Signatures was a concept disseminated by the German philosopher Jakob Böhme (1525-1624) that placed herbology on a religious footing. Böhme's idea was that God had left his signature(s) on specific herbal plants in the form of floral colours and shapes that indicated which body part or tissue benefited from the use of that herb. By reference to the appearance of numerous plants and flowers, the lung, the liver, the uterus and several other human tissues have all been implicated in this approach, and names such liverwort persist to this day. Although the foxglove plant itself has largely escaped reference linking its physical appearance and mode of action, the etymology of the name 'foxglove' has been a topic of much debate.

In contrast to the speculation surrounding the name foxglove and the folklore associated with the plant, Withering clearly based his research on digitalis on a firm scientific footing. Digitalis is derived from several cardiac glycosides produced by the plant and continues to be widely used as a heart medication. Extracts of the plant have been used to increase the strength of cardiac contractions. Digitalis acts to regulate heartbeat and is also employed as an antiarrhythmic agent to control the heart rate, particularly in individuals affected by irregular atrial fibrillation and especially if they have been diagnosed with congestive heart failure.

Oleander (*Nerium oleander; Thevetis peruviana*).

Oleander is a large evergreen shrub of the Apocynaceae (dogbane) family. It is also known as Nerium and is generally assumed to originate from North Africa and the eastern Mediterranean region, although it also currently occurs in southern U.S. states such as Florida and parts of Texas. The flowers appear as pink, red or white and often feature in ornamental parks and gardens where they are sometimes mistaken for rhododendron and other similar plants. As an illustration of this confusion another member of the dogbane family, the yellow dogbane *Thevetis peruviana* has also been given the name Oleander. The leaves and roots of Nerium, have a long record as a remedy for a wide variety of disorders including heart disease, asthma, diabetes and epilepsy, as well as infectious maladies such as malaria, ringworm and venereal diseases.

In addition to its pharmacological uses, Oleander extract also has a rich mythical history as a mood enhancer and aphrodisiac, which stretches back to ancient Greece. There are several accounts of its role in 'romantic enhancement' caused by smoking Oleander leaves or by drinking wine laced with the plant extract. However, some of these stories have been met by varying degrees of scepticism, usually resulting from differences in plant identification in the mists of time. Other plants assumed to be Oleander include Sweet Bay and laurel, although confusion also arises from its alternative name Rose Bay. Even within the Nerium genus, differences in flower colour often carry stark differences in assumed mythical characteristics: red for danger; pink for passion; white for innocence and purity. Not surprisingly, pink Oleander flowers in particular remain a popular part of romantic folklore.

Oh Leander!

The name 'Oleander' is thought to derive from the mythical character Leander who is immortalised in the ancient Greek story of Leander and Hero. According to legend these two lovers were separated by Hellespont, which now translates as the Strait of Gallipoli or the Bosphorus. They were only able to meet when Leander swam across the strait to be with his sweetheart, Hero, a priestess of Aphrodite, the Greek goddess of love. Hero would guide Leander in his crossing of Hellespont by lighting a torch. When this beacon was extinguished during a violent storm Leander lost his way and was drowned. Hero's desperate cries "Oh Leander" became synonymous with the plant that bears her lover's name.

The story of Hero and Leander has since appeared in poems by the 6th century Greek poet Musaeus Grammaticus, James Henry Hunt (1819), Christopher Marlowe (finished after Marlowe's death by George Chapman) (1821) as well as an epigram by Jonne Donne (1896). Further references to the myth of Hero and Leander appear in many languages, and it is difficult to find a myth that has been the subject of more prose and poetry, including works by Keats, Byron and Shakespeare.

In Shakespeare's *Much Ado About Nothing* (Act 5, Scene 2) Benedick talks of

> *"---The god of love*
>
> *That sits above*
>
> *And knows me, and knows me,*
>
> *How pitiful I deserve –*
>
> *I mean in singing. But in loving, Leander the good swimmer ---"*

The story also extends into classical artwork through paintings of Hero and Leander by Peter Paul Rubens (1604) and J.M.W. Turner (1837). This romantic narrative is further adorned with Paintings of Oleander flowers by Lawrence Alma Tadema (1882), Vincent van Gogh (1888) and Gustav Klimt (1892) .

This rich romantic mythology of Leander and the Oleander plant is blighted, however, by the fact that all parts of the plant are highly toxic, and that, following ingestion, there is only a very narrow margin between beneficial and lethal effects. The dark side of the Oleander story relates to the poisonous nature of all parts of the plant. Its alternative name 'dogbane' gives us a clue to its toxicity to dogs and other pets unfortunate enough to encounter it in gardens and on rural land. It is also toxic and even lethal to humans exposed to it accidentally or as a deliberate poison.

During one of his military campaigns Alexander the Great is reputed to have accidentally poisoned some of his troops by feeding them meat that had been roasted on skewers made from Oleander stems. Alexander himself is rumoured to have been deliberately poisoned by Oleander, although this may be another case of mistaken identity. There are many contradictory accounts of Alexander's death including White Hellebore *(Veratrum album)* as the lethal agent, not to mention advanced alcoholism and battle wounds.

Symptoms of Oleander poisoning include gastrointestinal discomfort, nausea, lethargy and dizziness. However, death mostly results from cardiac malfunction such as slowing of the heart rate and eventual heart failure. The primary toxic ingredient is the cardiac glycoside, oleandrin, consisting of four six-sided carbon rings and two five-sided rings plus numerous chemical side-groups.

Natural Toxins as medicines?
Perhaps not!

Alcohol.

Ethyl Alcohol (Ethanol).

When considering the psychoactive properties of Oleander and other plant extracts added to wine, we shouldn't forget the psychotropic effects of wine itself. Alcohol, primarily ethyl alcohol (C_2H_5OH), is a natural toxin produced by the fermentation of a large array of fruit and other plant products in conjunction with natural yeasts together with varieties of yeast that are cultivated specifically to aid the fermentation process. Although grapes are, of course, the favoured ingredient of wine, the name 'wine' has been used to describe beverages fermented from a large array of fruit and vegetables, even root crops such as potatoes. Evidence of fermented drink made from rice, honey and several kinds of fruit can be traced back as far as 7000 BCE in China. Although rice wine retains its name today, the involvement of cereals in the fermentation process, tends to shift the name of the end product from wine to beer, with terms such as 'barley wine' somewhere in the middle.

Both wine and beer-making have long histories. Winemaking can be traced back several millennia to the Caucasus region between the Black Sea and Caspian Seas. In the country of Georgia archaeological clues point to what may be the earliest known evidence of winemaking from grapes, dating from 6000 BCE. Further south, in Armenia, close to the Persian border (modern day Iran) an extensive cave system revealed comprehensive evidence of 6000-year-old winemaking activity including both the pressing of grapes and distillation of the product.

Many might argue that the name 'natural toxin' might seem too strong to apply to alcohol, particularly in view of its reputation as probably the world's most popular inebriant. However, the word 'intoxication' remains the most common term used to describe the effect of alcohol following its ingestion. Mild alcohol intoxication results in a feeling of well-being, although at higher levels of alcohol consumption this can give way to increasingly uninhibited and irrational behaviour that can result in harm or injury.

The natural fermentation of plant sugars by yeast is a self-regulating process that is limited by the concentration of ethyl alcohol that can be reached before the yeast is killed by the alcohol. To find this maximum level tolerated by the yeast, we need to look no further than the labels on wine bottles. Most commercial wines show an alcohol concentration between 12-15%, which reflects the upper level for alcohol that is tolerated by the yeast. Higher alcohol levels require a man-made intervention known as distillation. This boost to the alcohol content takes advantage of the fact that the boiling point of ethyl alcohol is about 78.4°C, compared with 100°C for water. Controlled heating of wine to the boiling point of alcohol vaporizes the alcohol enabling its capture in purified form by a condenser. Some water and other solutes are carried over with the alcohol during the distillation process, with the result that the final alcohol content of the distillate is in the region of 95%. This is normally diluted to about 40% alcohol to create products such as brandy (from grape wine) and whiskey (from beer) that are for safe human consumption. It is also common practice to add wine distillate back to a wine to create hybrid 'fortified wines', commonly known as port and sherry.

Although ethyl alcohol, ethanol, is by far the most common end product of the fermentation process, other alcohols are also generated. Methyl alcohol, commonly known as methanol (CH_3OH) is another fermentation product resulting from the hydrolysis of pectins which are naturally present in many fruit. Pectins are polysaccharides that are broken down by pectinase enzymes as part of the ripening process. Although methanol is a by-product of this process, it is not usually produced in toxic quantities as a result of natural fermentation. However, due to its use as a fuel and an antifreeze agent, availability of methyl alcohol in lethal concentrations may, occasionally, result in accidental poisonings. Such cases may be aided by mistaking methyl alcohol for the less toxic ethyl compound. Butanol (C_4H_9OH) can result from bacterial fermentation but not generally in toxic concentrations. High levels can cause delirium, loss of consciousness if ingested, although this is rare.

Although a dominant theme in the history of plant-based drinks has been the production and manipulation of alcohol content by fermentation and distillation, other additives serve no other purpose than to improve the palatability of the product. A typical example has been the addition of hops to beer, with no motive other than to provide a popular, characteristic flavour. As we have seen in the case of oleander, however, some additives go further than supplying extra flavour!

Wormwood.

The common wormwood *Artemisia absinthium,* is one of over a hundred species of the *Artemisia* genus belonging to the daisy or aster family, Asteraceae. The wormwood plant is widespread in temperate European, America and Asian countries as a perennial shrub common in hedgerows and rough pasture. Despite its unremarkable appearance, it has a rich history including a notable reputation as a folk medicine. Clues to its medicinal properties are found in its nomenclature and related mythology. The generic name *Artemisia* is derived from Artemis, the Greek goddess of the hunt, of nature and, interestingly, of childbirth. The plant has a long history stretching back to the Chinese emperor Shen-nung (2800 BCE) who has been credited with the discovery of wormwood as a cure for malaria. Its reputation as a treatment for parasitic and other intestinal disorders continued in ancient Greece where the medicinal uses of wormwood were mentioned in Ebers papyrus ca. 1550 BCE and the writings of Hippocrates in the third and fourth centuries BCE. These have since led to its remedial use for a broad spectrum of medical conditions including gastritis, hepatitis and osteoarthritis. Relief from gastrointestinal, menstrual and obstetric pain by *Artemisia* probably help explain the allusion to Artemis as the childbirth goddess.

The etymology of the Early English word for wormwood, 'Wermod', has led to its identification as Vermouth, a popular bitter-tasting additive to a variety of alcoholic beverages including wine and gin. The medicinal properties of vermouth predate its use as an aperitif, however, and again take us back to Ancient Greece and the promotion of vermouth as an analgesic, antioxidant and anti-inflammatory agent. Other more contemporary medical applications of wormwood extract point to its anti-ulcer, anticarcinogenic, hepatoprotective and cytotoxic activity.

Despite historical accounts of the medical attributes and mood-altering effects of wormwood extracts, there remained serious concerns over neurotoxicity associated with anything but very small doses. A significant part of this concern related specifically to a highly alcoholic infusion of wormwood, fennel, anise and other leaves and flowers in an herbal distillate with an alcohol content in the 35-75% range. The resulting, green-coloured spirit became known as absinthe, following the species name of the common wormwood, *A.absinthium*. Although originating in Switzerland in the late 18th century, absinthe gained in popularity in other European countries throughout the 19th century. The attraction was both medicinal and recreational, leading to mixed signals surrounding its use. During efforts by the French military to colonize parts of North Africa in the 1830s absinthe became a popular addition to water from hot desert climates. Although this use of absinthe was ostensibly as a vermicide and antibiotic, the mind-altering properties did not go unnoticed by the artistic community where it became known as "The Green Fairy". In Paris by the turn of the 20th century, absinthe had become the preferred drink of the artistic elite, spurred on by its reputation as a wellspring of creativity, not to mention an aphrodisiac! Enthusiastic absinthe drinkers included Charles Baudelaire, Lewis Carroll, Edgar Degas, Alfred Jarry, James Joyce, Guy de Maupassant, Arthur Rimbaud, Henri de Toulouse-Lautrec, Oscar Wilde and Vincent van Gogh. The consequences were not always positive however. Absinthe has been blamed for contributing to the deaths of Baudelaire and Jarry and for the self-inflicted loss of van Gogh's ear. Ernest Hemingway, another enthusiastic Green Fairy drinker, described it as

"opaque, bitter, tongue-numbing, brain-warming, stomach-warming, idea-changing liquid alchemy, ----- It's supposed to rot your brain out, but I don't believe it. It only changes the ideas.".

The non-literary establishment were largely unconvinced.

Concerns over the neurotoxic effects of absinthe led to its ban in Belgium, France. Italy, Netherlands, Switzerland and the United States in the first half of the 20th century, although these bans were relaxed beginning in the 1990s, when it was decided that the primary cause of absinthe toxicity was its high alcohol content. Outright bans have since been largely replaced by strict rules and warnings governing dosage and ingredients. Long-term use of *A. absinthium* oil may cause toxic and mental disorders in humans with clinical manifestations including convulsions, sleeplessness, and hallucinations.

Aside from the varied medicinal uses of wormwood extracts and concoctions, it is, perhaps, the antifungal, antimicrobial and antiviral properties of plant oils derived from wormwood that are the most powerful reminders that the compounds responsible for all this bioactivity are natural toxins that perform protective roles for the wormwood plant. Major bioactive components of wormwood include the monoterpenes α- pinene, β-pinene (see Table 1) which contribute to the aromatic odour and bitter taste of wormwood. However, probably the most toxic ingredients are the monoterpenes α- and β-thujone. The neurotoxicity of the thujone molecule arises from its function as antagonist of the GABA receptor on neural membranes (see Table 2). This results in an uncontrolled depolarisation of neural membranes resulting in seizures and convulsions.

Both the bitter flavour of the wormwood and the lethal consequences of its ingestion are cited in the biblical Book of Revelation, Chapter 8, verses 10 and 11:

10 The third angel sounded his trumpet, and a great star, blazing like a torch, fell from the sky on a third of the rivers and on the springs of water—

11 the name of the star is Wormwood. A third of the waters turned bitter, and many people died from the waters that had become bitter.

Although this reference has been subjected to numerous variations in both text and interpretation it clearly illustrates that wormwood was firmly embedded in GrecoRoman mythology at the time the Book of Revelation was written 95-96 CE.

Juniper.

The juniper 'berry' is not a fruit, but a female seed cone belonging to members of the coniferous family of trees and shrubs known as the Cupressaceae. The discovery of juniper berries in several ancient Egyptian tombs, probably reflects their long history of remedial and medicinal use together with supposed mystical powers that extended into the afterlife. The perceived importance of these herbs is reinforced by the fact that they are non-native to Egypt and were presumably imported for their spiritual powers. Where juniper trees are found in their native environment, such as parts of Europe and North America, they have often been accompanied by a rich folklore including real or imagined remedial properties such as the promotion of increased strength and wellbeing. The North American indigenous Blackfoot community used juniper berry tea as an anti-emetic.

Juniper extracts have been used as cleansers and disinfectants for both internal and external use including aromatherapy. For example, the North American Crow tribe recorded the use of juniper extract as a bactericidal drink to accompany childbirth. However, in some cultures the juniper berry is regarded as a form of birth control. In some instances, the reputation of the juniper tree extends to more supernatural powers such as the linkage of its chopping down with a run of bad luck for those responsible. In addition to their proven remedial properties, probably the best-known use of the juniper berry is its distinct flavouring of a gin and tonic, featuring the characteristic flavonoid, juniperin oil plus a diverse cocktail of fatty acids, resins, terpenes and aromatic compounds such as pinene and myrcene. The drink famously incorporates the presumed remedial properties of the juniper berry with its alcoholic base.

Nicotine.

Nicotine (Figure 26.) is a natural alkaloid found in plants of the nightshade (*Solenaceae*) family e.g. tobacco (*Nicotiana*) and Pitchuri thornapple *Duboisia hopwoodii,* potato, tomato, eggplant. It is produced in plant roots and is transported to the leaves where it acts a powerful deterrent against herbivorous insects. Nicotine from the root system of some plants may also deter the germination and growth of some other nearby, potentially competing plants, a so-called allelopathic effect.

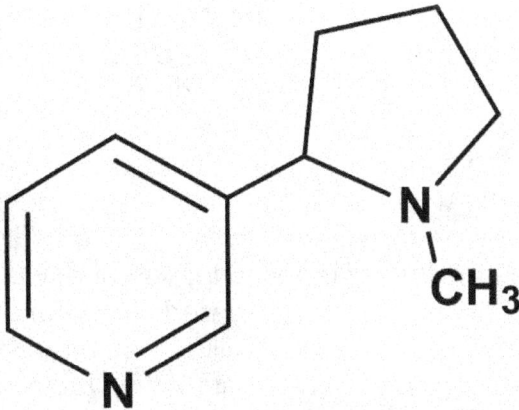

FIGURE 26 Nicotine

Despite the potent insecticidal properties of Nicotine, some insects are known for their immunity to its toxic effects and can actually benefit from the ingestion of plants containing high levels of the alkaloid. Among such 'specialist consumers' perhaps the best known is hornworm caterpillar *Manduca* which feeds on several different members of the *Solenaceae* including tobacco, tomato, eggplant and peppers. The caterpillar uses the enzyme CYP6B46 to facilitate the transfer of nicotine from its midgut to its haemolymph (blood) from which it is released into the external environment through the respiratory openings (spiracles) that line the body of the larva. The hornworm benefits by 'borrowing' the nicotine from the ingested plants to fend off predatory insects and spiders.

The earliest sign of human tobacco use consists of tobacco plant seeds retrieved from a fireplace along with other artefacts in a 12,300-year-old settlement in the Great Salt Lake Desert area of northern Utah in the U.S.A. This discovery extended the previous evidence of tobacco use by about 9000 years, although there is patchy information establishing its probable origin and likely cultivation in north and south America going back to the Pleistocene era. Its introduction into Europe accompanied several explorers who were intent on spreading their influence from the 'Old World' to the 'New', mainly from the early 16th century onwards. However, on the other side of the world, in Australia, tobacco has been used by native populations for several millennia, primarily in the form of a chewable mixture of tobacco leaves and wood ash known as pituri. Pituri was mainly prepared from the tobacco plant *Duboisia hopwoodii*, and fulfilled several functions in Australian aboriginal society, where it was used as a stimulant in lower doses but was capable of inducing a more hallucinogenic state in larger quantities. Pituri also had a hierarchical role in that its usage was often confined to societal elders whose trance-like state was interpreted as spiritual enlightenment.

Nicotine was first introduced to Europe in the mid 16th century, when smoking tobacco was favoured as a means of protection against various illnesses including the plague. It was first isolated from the tobacco plant in 1928 by the German chemists Wilhelm Possert and Karl Reimann, although its chemical structure [1-methyl-2- (3-pyridyl-pyrrolidine), $C_{10}H_{14}N_2$] was first proposed in 1892 and was confirmed by synthesis in 1895.

The mood enhancing properties of nicotine that are responsible for its global recreational use result from its binding to nicotinic receptors in the endocrine gland known as the adrenal medulla. Binding to these receptors causes cell depolarization and an influx of calcium through voltage-gated calcium channels. Calcium triggers the synaptic release of the hormones dopamine, epinephrine (adrenalin), norepinephrine and acetyl choline into the bloodstream, resulting in an increased heart rate, blood pressure and respiration, as well as higher blood glucose levels.

Epibatidine.

Epibatidine is a heterocyclic amine compound, also defined as a chlorinated alkaloid, similar in structure to nicotine. It was discovered by John W. Daly in 1974, who reported it as a medically important natural skin product of the Ecuadorian frog *Epipedobates anthonyi*. Despite some debate about its definition as a truly natural product based on the very small amount available from this specific natural source of the compound its structure was not confirmed until 1992. Epibatidine (from Ecuadorian frog) is a heterocyclic amine compound, similar in structure to nicotine and acts as a receptor agonist at most nicotinic acetylcholine receptors In low doses it is a highly effective analgesic (more than 200x as effective as morphine). The discovery of epibatidine was attributed to John Daly in 1974, although some initial doubts were raised as to the original natural source of the compound its structure was not confirmed until 1992. However, unlike many analgesics such as morphine, epibatidine binds to either muscarinic or nicotinic acetylcholine receptors (nAChR) rather than opiate receptors.

Strychnine.

Strychnine (figure 27) is highly toxic indole terpene alkaloid most commonly found in the seeds and bark of members of the *Strychos* family of plants, most commonly *Strychnos nux-vomica*, a tree native to tropical forests in Southern India, Sri Lanka, Philippines, Indonesia and other parts of eastern Asia.

It was first discovered by French chemists Joseph Bienaimé Caventou and Pierre-Joseph Pelletier in 1818 in the Saint-Ignatius' bean *Strychnos ignatia*. The structure of strychnine was first determined by Robert Robinson in 1946 and in 1954 this alkaloid was synthesized in a laboratory by Robert B. Woodward, feats that won both of these chemists Nobel prizes.

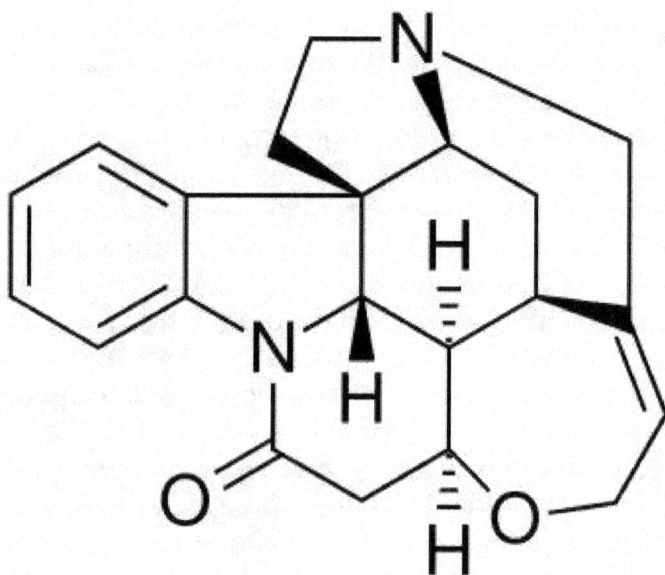

Figure 28. Strychnine.

Few compounds have matched the case history of strychnine as a poison. However, it is perhaps surprising that such a notorious poison would initially have found fame as a tonic with a reputation as a mood enhancer and as a remedy for several ailments. As recently as the late nineteenth and early twentieth centuries strychnine was marketed as a heart and bowel stimulant with the ability to strengthen skeletal musculature. Therapeutic use of strychnine persisted even though its dual nature as both a medicine and a poison had long been a matter of record in its countries of origin. In recent decades the balance has shifted away from medical use of strychnine which is now universally prohibited because of its toxicity. Nevertheless, strychnine retains its poisonous reputation through its continued use in some countries as a means of killing rodents and other predators and pests. It is, perhaps somewhat ironic that this market continues to drive the availability of strychnine, which has led to its occasional use as an agent of murder and suicide. While cases of murder by strychnine appear from time to time, it should be added that several of these stories of murder by strychnine are fictional in origin, with narrators as varied as Arthur Conan Doyle (*The Sign of Four*), Agatha Christie (*The Mysterious Affair at Styles*) and Alfred Hitchcock (*Psycho*). One reason for the comparatively rare use of strychnine as a deliberate poison is the highly characteristic nature of the symptoms accompanying its ingestion. Unlike a toxin such as arsenic where fatal symptoms may be indistinguishable from some ailments, the often-dramatic results of strychnine poisoning are difficult to disguise and may provide information on the source compound and its characteristic physiological activity.

Several cases of strychnine poisoning, both real and fictional, provide clues relating to the alkaloid's mode of action. Strychnine is a powerful neurotoxin which is a competitive antagonist of the inhibitory neurotransmitter glycine. To understand the effect of strychnine it is first necessary to comprehend the action of the amino acid glycine in its role as a regulator of neural impulses. When associated with its designated (glycine) reactor, glycine inhibits the conductance of neural impulses in the nervous tissue of the spinal cord and brain. When activated by glycine, the (glycine) reactor, through its control of gated ionic (calcium) channels in the neural membrane, raises the threshold for the number of excitatory neurotransmitters required to generate an action potential. This exerts a 'braking effect' on the motor nerve fibres in the spinal cord which control muscle contraction.

The inhibitory effect of glycine is reversed by the antagonistic action of strychnine. Through its role as an antagonist of glycine, strychnine forms a covalent bond with the glycine receptor, preventing the inhibitory effects of glycine on the post synaptic neuron. This causes a flood of negatively charged chloride ions into the neuron via the chloride channel. The change in polarity lowers the threshold for triggering action potentials which become more easily activated. At higher strychnine doses this may result in muscular convulsions and eventual death by asphyxiation. This sequence of physiological activity also provides a rationale for the historical use of strychnine in small doses as a muscle stimulant in heart, bowel and skeletal muscle. At the same time it provides yet another object lesson in the importance of dose in differentiating between remedial and lethal effects.

Fungal Toxins.

Estimates of existing fungal species vary widely, usually between 1.5 and 5 million, although only about 150,000 have been positively identified.

Part of the reason for this relates to the fact that most members of this group of plant-like organisms are hiding in plain sight and have been doing so since most living organisms first appeared on earth. It now seems likely that fungi were derived from ancestors whose fossil remains can be dated to at least a billion years ago. A critically important aspect of fungal evolution was the development of filamentous structures called hyphae which collectively formed masses known as mycelia. These fungal mycelia formed a symbiotic association with green algae which hitherto had inhabited a completely aquatic environment. The structure formed from result of this fungal/algal association became known as the mycorrhiza, which evolved as an integral part most plants, i.e. the root system. Mycorrhizal tissue provided a much-expanded surface area that allowed plants to derive water and nutrients from soil. Today more than 90% of all plants owe their terrestrial existence to the evolution of this ancient mycorrhizal/algal hybrid.

Today perhaps 5% of fungi are regarded as edible, with only a tiny percentage falling into the "toxic" category. The vagueness associated with some of these numbers is perhaps summed up by the title of Terry Pratchett's book, *"All Fungi Are Edible, Some Only Once"*.

Fungi comprise several different taxonomic groups that would not naturally be considered as edible, but that are regularly ingested as a result of their contamination of food items that form part of the regular diet of both humans and their livestock, another consequence of their ubiquitous nature. Common examples are rusts and moulds. Yeasts, too, are common components of human food, either accidentally or deliberately, and are universally added to bakery products and drinks as an essential ingredient of the fermentation process. Moulds include several of the most toxic fungal species, such as *Aspergillus*, *Fusarium*, *Mucoromycetes* and *Scedosporium*. However, the fungi responsible for by far the most human deaths, as many as 95%, belong to mushrooms of the genus *Amanita*. Of these the most notorious is the aptly named Deathcap *Amanita phalloides*. The death toll resulting from eating Deathcap mushrooms includes the Roman emperor Claudius in the year AD 54 who is reputed to have succumbed to this species of mushroom supplied by his wife Agrippina. Voltaire claimed that Holy Roman Emperor Charles VI may have been a victim of accidental poisoning by Deathcap mushrooms in 1740, with subsequent historical ramifications leading to the Austrian War of Succession.

Amatoxins.

The principal natural toxins in *Amanita phalloides* are the eponymous amatoxins These are polypeptides essentially composed of 8 amino acids with a cyclic nitrogen-containing core. They are sometimes described as alkaloids due to presence of nitrogen in the cyclic core structure. Perhaps the most toxic component of *Amanita* mushroom is α-amantin ($C_{39}H_{54}N_{10}O_{14}S$). α-amantin is potent inhibitor of RNA polymerases II and III which are responsible for the transcription of DNA to messenger RNA precursors. Inhibition of these polymerases essentially destroys protein synthesis with deadly consequences.

Trichothecenes.

Trichothecenes are a large, diverse group of <u>mycotoxins</u> produced by various species of fungi, notably the mould *Fusarium*. Their core cyclic configuration includes a highly reactive epoxide structure wherein an oxygen molecule is linked to two adjacent carbon atoms at the 12,13 carbon positions, as well as a double bond at the 9, 10 carbon positions. Examples of fungal trichothecenes include Satratoxin and Roridin. These two functional groups are primarily responsible for trichothecene ability to inhibit protein synthesis resulting in a variety of cytotoxic effects including apoptosis[16] and the inhibition of RNA and DNA synthesis. The most toxic member of the trichothecene family of compounds, the so-called T-2 toxin, has a variety of adverse effects on mammalian consumers including hepatotoxicity, gastrointestinal toxicity and cardiovascular failure at high doses.

Perhaps the most notorious narrative involving the tricothecene T-2 toxin was its rumoured use by the Soviet Union as a chemical weapon in Laos in 1975-1981. This accusation, which originated from the office of the U.S. Secretary of State, was embellished by reports of toxic "yellow rain" delivered as a spray from aircraft. A more straightforward interpretation of trichothecene poisoning from this era focussed on the consumption by affected personnel of food contaminated by fungal products. As for the yellow rain portion of the narrative, this was attributed to clouds of faecal droppings from jungle bees that were given a yellow tinge by

---

[16] In normal metabolism apoptosis is a process whereby an organism eliminates unwanted cells in a controlled manner. It can be used to get rid of damaged or cancerous cells without causing inflammation. However, excessive apoptosis of neurones can lead to neurodegenerative diseases such as Alzheimer's and Huntington's disease.

the presence of pollen. Whatever the truth behind the military use of T2-toxin, a similar narrative involving its purported use as a bio-warfare agent appeared during the Gulf War in 1991. In this version, mycotoxicosis was diagnosed in victims of an aerial missile attack resulting in dermal blistering and skin necrosis. Symptoms of gastrointestinal exposure were reported to be nausea, vomiting and diarrhoea.

Psilocybin.

Psilocybin is a naturally occurring phosphoryl compound with a dimethyl core, containing nitrogen both as part of its cyclic structure and associated amine group, a structural characteristic that also justifies its definition as an alkaloid. Although its most common chemical name is 4-Phosphoryl-N,N-dimethyltryptamine (4-PO-N,N-DMT) it is still usually identified as psylocibin which is an hallucinogen commonly associated with some species of mushroom. Clues from cave paintings and other artefacts point to the use of hallucinogenic mushrooms as part of ancient Mediterranean culture. The exact timescale of the appearance of 'Magic Mushrooms' in the New World remains uncertain but seems to predate 16th century accounts by Spanish explorers of their ritual use in Central America.

One of the most puzzling attributes of fungal psilocybin relates to a specific aspect of its hallucinatory role; namely the ability to turn a potential consumer into an irrational zombie. Examples include zombie ants and flies that exist in a state of semi-paralysis following exposure to psilocybin. The most obvious interpretation of the hypnotic action of psilocybin is that of a chemical defence against a would-be predator. There are many defensive ploys that may be used against a consumer including lethal or sublethal toxicity or simple chemical repellence. However psilocybin seems to take this a step further by creating illogical behavioural patterns that are difficult to interpret in purely defensive terms. Perhaps its closest resemblance to a deterrent strategy is simply the infliction of a behaviour pattern on the potential insect predator that distracts it from the task in hand, i.e. eating the fungus! It is also likely that the subjugation of a potential consumer such as an insect may signify predatory behaviour by the fungus itself. An example might be a change in insect behaviour that makes it more vulnerable to colonisation by fungal spores or mycelia. In this way fungi are often seen as parasitic in their relationship with insects and even other fungi.

Cordyceps is a large genus of parasitic fungi comprising over 400 species which were originally classified in the family Clavitcipitaceae. However, the taxonomy continues to be revised to accommodate variants of a complex phylogeny that continues to be influenced according to a spectrum of host organism(s) and environmental conditions. Members of the *Cordyceps* genus are distributed among all continents in tropical and sub-tropical regions.

Although the bioactivity of natural toxins produced by *Cordyceps* seems clearly related to its parasitic *modus operandum* the pharmacological properties of *Cordyceps* have long been recognised and its use as a medication in China goes back several hundred years. It was first introduced into Europe in the late 17th century. The fungal source of medicinal agents was identified as *Cordyceps sinensis*, which together with *C. militaris* continues to be used for a broad range of pharmacological activities. These include maintenance of liver and kidney function as well as having anti-inflammatory, antitumour and antioxidant activity.

There are several chemicals derived from *Cordyceps* that are beneficial to consumers in a variety of ways. Notable among these are polysaccharides and sterols which have a wide range of pharmacological effects ranging from anticancer, and immunoactivity. Other biologically active compounds found in *Cordyceps* include nucleosides and their phosphorylated analogues, nucleotides, that are necessary for the transfer of genetic information by virtue of their roles as precursors of RNA and DNA.

Cordycepin (Figure 28) is a derivative of the nucleoside adenosine with a range of therapeutic properties. The range of effects is probably facilitated by its ready access to numerous intracellular sites due to its similarity to adenosine. In fact, cordycepin is identical to adenosine except for the absence of an oxygen atom at the 3' position of the molecule (in parentheses, see figure 28). Hence its chemical name 3'-deoxyadenosine.

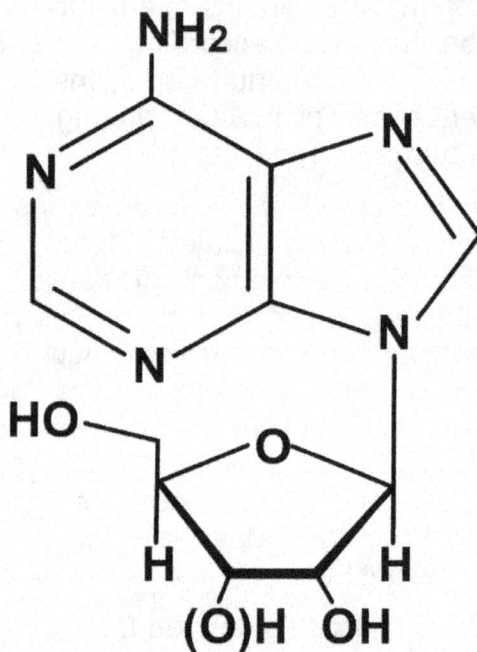

Figure 27. Cordycepin. This active component of Cordyceps differs from the nucleoside adenosine by the removal of the oxygen atom (in parentheses) from the 3' position.

Ergot Alkaloids.

Ergot fungi are a group of fungi belonging to the genus *Claviceps*, a group of about fifty species, most notably the species *Claviceps africana* and *C. purpurea*. These fungi grow on serials such as sorghum and ryegrass, leading to accidental, consumption by both humans and their livestock, often with deleterious results. Ingestion of ergot fungi has led to a broad range of ailments collectively known as ergotism. Symptoms vary greatly and have given rise to several subcategories of ergotism such as convulsive ergotism, enteroergotism and gangrenous ergotism. Convulsive ergotism derives from the ability of ergot alkaloids to modify the action of neurotransmitters such as dopamine, norepinephrine and serotonin causing muscle spasms. Ergot alkaloids responsible for these effects have a variety of names such as ergotamine, ergoclavine and ergometrine, dependent on variations of their multicyclic structure. Their ability to modulate and sometimes mimic the action of neurotransmitters no doubt stems from their shared structural similarities as derivatives of lysergic acid. Structural similarities with lysergic acid are also responsible for hallucinogenic properties of some ergot alkaloids. The psychedelic drug Lysergic Acid Diethylamide (LSD) discovered in 1938 is a derivative of *Claviceps purpurea*.

Slight variations in chemical composition of ergot alkaloids have resulted in a broad range of human pharmacological effects associated with accidental *Claviceps* consumption. In severe cases of ergotism reported as early as the 16th century, there are reports of vasoconstriction developing into gangrenous conditions leading to loss of limbs, and spontaneous abortion caused by severe uterine contractions. Other adverse symptoms of ergotism included an intense burning sensation of the skin that came to be known as St. Anthony's (or Holy) Fire, tremors, insomnia, excessive salivation, vomiting, and occasionally death.

The earliest documented case of ergotism occurred in AD 857 in the Rhine Valley. Multiple cases of human ergotism have been reported from Europe and Asia over the last 500 years, including reports linking convulsive and hallucinogenic symptoms to cases of bewitchment. These have included a popular but disputed theory that an outbreak of ergotism may have been at the root of the Salem Witch trials in Massachusetts, U.S.A. in the late 17th century.

The link between ergotism and fungal infection of grain such as rye was first made by the French physician Denis Dodart in 1676. This prompted a better recognition of the cause of the disease and a slow but often faltering initiative for the better management of crops prone to contamination by *Claviceps*. Although this has reduced the threat of human ergotism over several decades the disease remains a concern for farmyard animals fed contaminated grain products (Bennett and Klich, 2003).

A positive development in recent decades, however, has been an increased initiative to harness the therapeutic properties of ergot alkaloids. As is the case with several natural toxins, the history of ergot alkaloids often illustrates a delicate balance between their potentially harmful effects at higher concentrations but remedial uses at much lower doses.

This balance is illustrated by the hazard of premature abortion as a historical consequence of ergot ingestion, compared with its use as a childbirth aid at low doses of ergot alkaloids, a therapeutic function recognised by the German physician Adam Lonitzer as early as 1582.

Therapeutic properties of ergot alkaloids have now been recognised for a broad range of maladies, including gastrointestinal, reproductive and neurological disorders, with the prospect that fungal genomes may be manipulated to generate strains capable of producing therapeutic alkaloids with applications for medicine and agriculture (Panaccione 2023).

Cyanotoxins.

Sometimes confusion arises in the literature as a result of a common derivative word that leads to divergent meanings. An example is the term 'Cyanotoxins'. We have already seen how the nomenclature of the cyanide chemical group is derived from its Prussian Blue ('Cyan') ancestry. At first sight we might, therefore, expect that 'Cyanotoxins' are somehow related to the cyanide chemical group. Not so! The term cyanotoxins as used in this context relates to an unrelated mixture of chemicals having no relationship other than a common biological source, namely cyanobacteria. Further confusion might arise from the singular ability of this group of bacteria to photosynthesise, thus earning the name Blue-Green algae despite their simultaneous definition as bacteria. It is simply the blue-green colour of this taxonomic group that forms the basis for the word 'cyanotoxin(s)' used to describe the specific group of natural toxins that are characteristic of the cyanobacteria. The name 'cyanotoxins' therefore has nothing to do with the cyanide group of chemical compounds and simply refers to the source organisms for this mixed group of natural toxins, blue-green algae. The descriptive term 'cyanotoxins' is therefore used in the same way that 'fungal toxins' relate to their biological source, i.e. fungi.

Despite the absence of the deadly cyanide ion, blue green algae are highly toxic to a wide range of consumers. These include both primary consumers that filter-feed on the cyanobacteria themselves as well as numerous vertebrate and secondary consumers higher in the food chain.

The chemicals responsible for the highly toxic nature of cyanobacteria are a mixture of organic compounds dominated by cyclic peptides and alkaloids. Many are liver toxins (hepatotoxins) and neurotoxins. For example, microcystins and nodularins are cyclic hepatotoxins with a core structure derived from several amino acids. The high nitrogen content of these compounds is aided by the nitrogen-fixing capability of cyanobacteria. Cylindrospermopsin (CYN) is a cyclic guanidine alkaloid capable of blocking protein synthesis. In addition to their hepatotoxicity many of these cyanotoxins are inhibitors of protein phosphatases. This consequent interference with protein chemistry often has serious adverse effects on humans and other animals that might consume water and food chain organisms contaminated with cyanobacteria. The poisoning of humans and domestic animals by cyanotoxins is globally widespread and occasionally results in fatalities.

A 1996 incidence of hepatotoxicity following drought conditions in Caruaru, Brazil, was ascribed to the ingestion of cyanobacteria caused by drinking contaminated lake water. Of the 126 people affected by this outbreak, 60 died. Similar outbreaks of cyanobacteria poisoning have been recorded throughout the globe. Often these are exacerbated by eutrophication resulting from increased nutrient run-off and rising coastal water temperatures. In 2014 a toxic bloom of cyanobacteria was responsible for a directive issued by the town of Toledo Ohio U.S.A. that prevented people drinking water from Lake Erie. Similar concerns over 'harmful algal' (cyanobacteria) blooms are increasingly linked to climate-related warmer temperatures in aquatic environments as diverse as freshwater ponds in India and parts of the Baltic Sea.

However, when cyanobacteria are examined on a geological time scale a completely different picture emerges, showing that these bacteria played a critical role in the creation of a relatively oxygen-rich environment more than two billion years ago[17], thereby pre-dating the evolution of metazoan organisms. The poisoning of mammals and other vertebrates by cyanobacteria has only been recognised as a fairly recent phenomenon, with no recorded cases earlier than the 1800s. Despite the critical role that cyanobacteria have played in determining the evolution of the earth's atmosphere over billions of years, much of the current focus remains on emerging issues related to the part(s) played by cyanotoxins in aquatic ecosystems.

The Great Oxidation Event.

It is now believed that cyanobacteria comprise an ancient prokaryotic family, of free-living bacteria, the Cyanophyceae, originating more than two billion years ago. As many as 2000 species distributed among 150 genera have since been ascribed to this family. In view of this ancient provenance, it is now theorised that, as a result of their photosynthetic activity, the Cyanophyceae, were probably responsible for creating the earth's first oxygen-based atmosphere beginning over two billion years ago. As a result of their photosynthetic capability, blue-green algae were able utilise solar energy to split the water molecule, releasing oxygen gas into the atmosphere. Simply put, this capability is responsible for why we currently breathe and respire oxygen. This 'Great Oxidation Event' led to the gradual displacement of methane by oxygen as a major atmospheric constituent and the development of a new range of multicellular life forms utilising aerobic rather than anaerobic

---

[17] This "Great Oxidation Event" , between 2.4 - 2.2 billion years ago, initiated the transformation of the earth's atmosphere from an oxygen-free environment to one in which the proportion of atmospheric oxygen began its increase to the current level of 20%

respiration as the primary driver of energy and growth. The relative increase in atmospheric oxygen was initially associated with a cooling trend involving large scale encroachment of ice over much of the earth's surface. However, as the new flora and fauna thrived in the increasingly oxygenated environment the upsurge in biological growth was accompanied by an increase in the emission of carbon dioxide into the atmosphere and a reversal of the cooling trend that continues to this day.

Much speculation has been focussed on the reconciliation of the ancient historic roles of these natural toxins with their current influence on aquatic ecosystems. Several researchers have questioned the primary role of cyanotoxins as a defense against consumers. Part of the counterargument against cyanotoxins being evolved in response to grazing pressure has to do with the sheer age of the cyanobacteria.

One question might be "why develop a defense against potential grazers that, several billions of years ago did not yet exist?". This line of speculation has been supported by molecular studies suggesting that the genetic templates responsible for encoding cyanotoxins such as Microcystin (MS) and Cylindrospermopsin (CYN) were present long before the appearance of the metazoan grazers approximately 1500 million years ago. (Murray et al. 2011). Of course, this is not to say that cyanobacteria may have developed natural toxins that predispose them to defense against metazoans and other consumers and competitors that might appear later in the evolutionary timescale. However, from an evolutionary standpoint if we assume that the initial appearance of cyanotoxins long pre-dates the evolution of potential consumers and competitors of cyanobacteria, it is reasonable to ask what other underlying selection pressure(s) might have driven the evolution of this potent mixture of cyanotoxins apart from grazing pressure? This subject has prompted much speculation, some of which has been summarized by Holland and Kinnear (2013).

Among the alternative roles for cyanotoxins that have been hypothesised for cyanotoxins are an ability to improve nutrient metabolism. For example, MS synthesis by cyanobacteria has been shown to increase under phosphate-limiting conditions. In some cases, CYN production by cyanobacteria has been shown to induce increased alkaline phosphatase (ALP) in neighbouring phytoplankton species, thereby increasing the total inorganic phosphate available to the total phytoplankton community. Similar phenomena can also be found in other algal species. The dinoflagellate *Alexandrium tamarense* exhibited a 4 to 6-fold increase in saxitoxin production under phosphate-limiting conditions.

Other proposed 'non-competitive' benefits from cyanotoxins have included a strengthening of the 'cellular architecture' of the blue-green and other algae through a phosphorylation-based signalling process responsible for cytoplasmic homeostasis and cellular communication. These properties can result more versatile protein chemistry with improved resistance to both long-term and short-term changes in environmental conditions such as climate change.

Saxitonin (STX)

Saxitonin is a potent neurotoxin found in marine organisms such as puffer fish. However, it is most commonly associated with blooms of algae including cyanobacteria and dinoflagellates known as red tides. These Harmful Algal Blooms (HABs) are responsible for paralytic shellfish poisoning caused by the ingestion filter-feeding bivalves containing high concentrations of toxic algae such as cyanobacteria. Saxitonin is one of the most potent neurotoxins known and can be a potent inhibitor of electrical conduction in neural cells of consumers, leading to paralysis in serious cases.

The earliest documented case of Saxitonin poisoning was caused by toxic dinoflagellate *Alexandrium* in the coastal waters of British Columbia in 1793 when five members of the explorer Georges Vancouver's crew were taken ill after eating contaminated shellfish; one died.

A much more recent case of poisoning by Saxitonin was recorded in the early 1970s when several people were taken ill following the consumption of mussels from the North Sea coast of Northumberland in North-East England. Numerous fish deaths and seabird deaths were also recorded from the same region during the same period. The culprit was identified as a bloom (also known as a "red tide") of the dinoflagellate *Gonyaulax* (Figure 29) which entered the food chain as a result of massive bioconcentration by filter-feeding mussels. and other bivalves from coastal waters.

Figure 29. Gonyaulax. A dinoflagellate responsible for poisoning of vertebrate consumers including humans through the ingestion of filter-feeding bivalve molluscs. Bioconcentration of the neurotoxin saxitoxin follows harmful algal blooms known as 'red tides'.

Part of the identification effort involved the use of a Hardy Plankton Recorder, invented by the marine biologist Alister Hardy (1896-1985). This ingenious piece of equipment helped track down the root cause of the sickness by doing exactly what it was designed to do. In one important respect, this plankton recorder resembles an old-fashioned camera; of the type that use a plastic film that spools from one reel to another. However, the plankton recorder is a much larger device than the camera! Instead of a reel of plastic film there is a plankton net that unwinds from one spool to another while being dragged through the water behind a boat. The speed of this unwinding/winding action feeding the plankton net between the two spools is controlled by the speed of the vessel. As the vessel towing the recorder is equipped with navigational aids the precise location of every batch of filtered organisms and the time they were caught can be calculated. Responding to the need to determine the existence of a potentially noxious algal bloom that coincided with the poisoning outbreak investigators were able to track and analyse plankton samples plankton recorders that were routinely deployed by commercial vessels crossing the North Sea between the U.K. and Scandinavia according to known timetables and routes.

Tetrodotoxin.

Tetrodotoxin is a potent neurotoxin found in a wide range of seafood such as puffer fish (a.k.a. "fugu") and blue ring octopus as well as numerous marine species such as bivalves, crabs and moon snails. It is also found in several other animals (e.g. bivalves, rough-skinned newts, crabs and moon snails). In some cases, it derives from exogenous sources including the Proteobacteria *Vibrio* and *Pseudomonas* and algal species such as dinoflagellates.

When ingested by humans, usually as a result of consuming

contaminated seafood, tetrodotoxin can cause paralysis by affecting the sodium ion transport in both the central and peripheral nervous systems. A low dose of tetrodotoxin produces tingling sensations and numbness around the mouth, fingers, and toes. Higher doses produce nausea, vomiting, respiratory failure, difficulty walking, extensive paralysis, and in severe cases hypertension and death.

One of the first modern accounts of seafood poisoning attributed to tetrodotoxin was recorded by Capt. James Cook in 1774 following consumption of new species of fish prepared for dinner. Cook writes "About 3-4 o'clock in the morning (after dinner) we were seized with the most extraordinary weakness in all of our limbs attended with numbness of sensation like that caused by exposing one's hands and feet to the fire after having been pinched much by frost ......". When the natives came aboard the following day, only then did they make the point that on no account should the fish (fugu) should be eaten whole. It is still a matter of debate whether the delay in warning these newly arrived strangers of the toxicity of fugu toxicity may have had a mischievous element.

Although the sale of Pufferfish for human consumption is banned in Europe, it remains a gastronomic delicacy in some cultures to this day, but one carrying a significant risk. Fugu bones have been unearthed in places of human habitation in Japan going back over 4000 years. However, about five deaths per year in Japan have been attributed to its continued use as seafood in that country. Understandably, much is made of the need to remove the liver, the primary source of tetrodotoxin in that species. However, it is also worth bearing in mind that tetrodotoxin is about a thousand times more toxic than potassium cyanide! Needless to say, food preparation of this particular dish is a well-practiced art!

Tetrodotoxin is a sodium channel blocker. It inhibits the firing of action potentials in neurons by binding to the voltage-gated sodium channels in nerve cell membranes and blocking the passage of sodium ions (responsible for the action potential) into the neuron; hence paralysis.

Toxins used as Agents for Terror and Assassination.

Organophosphorus (OP) compounds

While the functional groups of organophosphorus (OP) compounds are part of the biosphere and organic derivatives of the phosphate ion comprise many biologically important compounds, neurotoxic OP compounds are largely synthesised and are, therefore, peripheral to our definition of 'natural' toxins. Nevertheless they provide a potent example of a toxic action common to many organic toxins, natural and otherwise; namely the inhibition of cholinesterase. Compounds that have this property are known as anticholinesterases.

Anticholinesterases (cholinesterase inhibitors) are chemicals that prevent the breakdown of the neurotransmitter acetylcholine leading to a build-up of acetylcholine at neuronal junctions. These synapses number as many as a trillion in a single human! Anticholinesterase activity increases the amount of the acetylcholine in the synaptic cleft that can bind to muscarinic receptors, nicotinic receptors and others. Cholinesterase inhibitors have been used in the treatment of Alzheimer's disease. It is believed that decreased levels of acetyl choline in the brain can cause Alzheimer's disease.

Among cholinesterase inhibitors several organophosphorus compounds were developed as nerve agents before the second world war, but the threat of use by both sides proved to be a deterrent to their use in the Second World War (WWII). Following WWII the primary application of OP compounds was directed to their use as pesticides. Organophosphorus (OP) compounds form the basis of pesticides such as Parathion but need to be formulated such that they can be rapidly metabolised by non-target organisms. The use of OP pesticides greatly increased in the 1970s following the ban of organochlorine pesticides such as dieldrin and DDT. However, some OPs have more sinister applications. Sarin, VX and Novichok are examples of chemical warfare agents that have been implicated in murder and assassinations. Sarin is a highly potent OP nerve agent that was used in a 1995 terrorist attack on the Tokyo subway system that resulted in the acute deaths of at least 13 people and the severe injury of more than 50 others, some of whom later died. A Sarin gas attack on Halabja in 1988 by Iraqi leader Saddam Hussein resulted in the deaths of as many as 5000 mainly Kurdish people. Syrian president Bashar al-Assad was responsible for the use of Sarin as a chemical weapon in the Ghouta suburbs of Damascus, resulting in the deaths of several hundred people in 2013.

More modern, apparently targeted, examples of alleged use of toxic OP compounds include the suspected 2017 Novichok poisoning of Russian activist Alexei Navalny and the alleged 2017 assassination of North Korean Kim Jong-Nam through the dermal application of VX. VX is an agent of even greater toxicity than sarin due to its lower volatility and more long-term environmental persistence. Several OP nerve agents are considered to be more than 25X more potent than cyanide.

Ricin.

Another biotoxin used in political murder and terrorism is Ricin, which can be highly toxic to humans. Ricin belongs to a family of compounds commonly known as lectins, which are carbohydrate-binding proteins commonly found in legumes such as lentils and beans, notably kidney beans and castor beans. Most acute poisoning episodes in humans have resulted from the oral ingestion of castor beans, 5–20 of which can prove fatal to an adult. The presence of castor beans in Egyptian tombs as old as 4000 thousand years suggests that the medical and spiritual properties of Ricin and other lectins have been known for millennia. Remedial applications have long been part of traditional Chinese medicine and include the uses of castor bean extract as an anti-inflammatory and antiseptic in wound dressing. Its use in Chinese medicine as a treatment of parasitic worms reflects the pharmacological use recorded by the Greek physician Dioscorides (279-206 BCE) including its internal use as a diuretic and emetic. As with many natural toxins the difference between beneficial and lethal effect may depend on the amount ingested, the biological source and method of preparation. For example, lectins are also found in vegetables of the *Solanaceae* family such as tomatoes and potatoes, but only in non-toxic concentrations. Also, extensive cooking has been shown to lessen or eliminate the toxicity of many lectins.

Despite the versatility of sublethal Ricin as a treatment for a broad range of medical conditions spanning thousands of years, it remains best known as a potent toxin with a sinister history as a deliberate poison often with terrorist connotations. Ricin acts by inhibiting and destroying ribosomes which are molecular complexes associated with the cellular organelle known as the endoplasmic reticulum. Ribosomes are responsible for the genetic transcription of proteins, which control numerous bodily functions. Their elimination causes the normal turnover of proteins to grind to a halt. This breakdown in protein replacement results in the deterioration of bodily functions and a broad range of symptoms which slowly develop, hours to days following ricin exposure.

The delay between ricin exposure and the development of toxic symptoms can disguise cause and effect, a property that has been used to conceal several murders. Perhaps the most famous example was the assassination of Georgy Markov, a Bulgarian playwright and dissident killed by Bulgarian Secret Service aided by the Soviet KGB in London in 1978. Markov was stabbed by a pointed tip of an umbrella dosed with Ricin. Markov appeared to die several days later from 'Natural Causes'.

Ricin continues to appear from time to time in individual murder cases, including fictional ones such as Agatha Christie's 1924 short story *"The House of Lurking Death"*, where the difficulty in diagnosis of the toxin is an important part of the plot. Although more modern analytical techniques have done much to clarify the mode of action of Ricin and have significantly improved its chances of its detection.

Quinones. A Group of Many Colours (and uses).

Quinones consist of a benzene core on which two hydrogen atoms are replaced by two oxygen atoms, forming two carbonyl bonds. They are notable as pigments in a wide range of living organisms, and some are used as indicators of the presence of hydrogen ions (H^+) in unknown solutions. Their properties as pigments has led to their use in textile, tanning and cosmetic industries, although in high concentrations quinones can cause dermatitis and eye irritation. In the state of California in the U.S. quinone has been classified as a hazardous air pollutant since 1993. Many quinones are strong oxidizing agents and are used in a variety of industries. Some quinones are natural plant products that are toxic to a broad range of potential consumers and competitors. Their broad range of biological activities have led to a variety of uses including some medical applications.

Naphthoquinones.

Naphthoquinones are a group of plant-based derivatives of aromatic hydrocarbons identified by a double benzene ring structure. This distinguishes them from single ring benzoquinones and three ring anthraquinones.

Juglone is a naphthoquinone that is derived from the black walnut tree (*Juglans nigra*). It has an allelopathic function i.e. toxicity to 'competing' plants and is particularly toxic to members of the *Solanaceae* (peppers, potatoes etc.). Such a high degree of 'targeted' toxicity to that particular family of plants is perhaps ironic in view of the notoriety of the *Solanaceae* in producing a considerable arsenal of toxic

175

compounds of their own. An example of plant chemical warfare perhaps!

Naphthoquinones have also been shown to be toxic to many herbivorous insects. Fungal naphthoquinones are toxic to bacteria. This antibiotic activity of naphthoquinones reflects their role as potent inhibitors of electron transport, as disrupting agents for DNA and as producers of powerful oxidants leading to cell death. Interestingly, these same qualities have led to their development as herbal remedies for a variety of ailments including anti-cancer, anti-inflammatory and anti-parasitic applications.

The toxicity of naphthoquinones to a broad range of aquatic organisms including cyanobacteria (harmful algal blooms) and has also resulted in their use as a ballast water treatment in ships to control aquatic invasive species. Globally an estimated 12 billion tonnes of ballast water are transferred each year, each ship carrying several hundred liters to >100,000 tonnes of ballast water. Ballast water is the dominant vector for the transfer of non-indigenous aquatic species between coastal areas that may be many thousands of miles apart. This can cause irrevocable changes in the food web with diverse consequences such as the destruction of commercial fisheries, disease outbreaks due to harmful algal blooms and clogging of industrial intakes and outlets by exotic organisms. In the U.S. the annual costs of aquatic bioinvasions has been estimated to be approximately $10bn. In the U.K. the annual cost of invasive species to the U.K. national exchequer has been estimated to be £1.7bn. Once an invasive species such as the zebra mussel is established in the new environment, it is nearly impossible to eradicate. The cost of efforts to control zebra mussels (clogging sewage treatment plant intakes, storm water outfalls and other structures) have been conservatively estimated at $1bn. Tests aboard seagoing vessels showed naphthoquinones to be

176

highly effective in controlling invasive planktonic organisms in ballast water (Wright *et al.* 2007). Of those naphthoquinones tested plumbagin and juglone (Figure 30) were found be the most toxic although menadione (Figure 30) proved to be the most cost-effective. While not as toxic as juglone and plumbagin in most instances, its current bulk manufacture of menadione for clinical purposes and as an animal feed supplement (as vitamin K3) makes it much cheaper to produce than the other naphthoquinones. A positive attribute of naphthoquinones as ballast water treatment agents is their photolysis to a non-toxic form on discharge, after they have effectively performed their function in a dark ballast tank during transit.

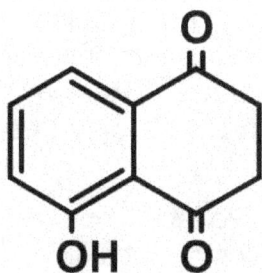

5-hydroxy 1,4-naphthoquinone (Juglone)	2-methyl 1,4-naphthoquinone (Menadione)

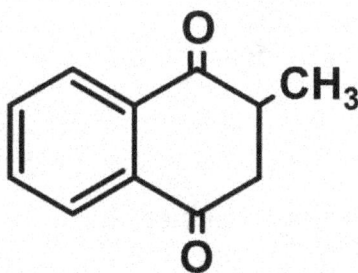

Figure 30. Naphthoquinones Juglone and Menadione. Examples of biologically active naphthoquinones with similar chemical structure but different properties Juglone is noted for allelopathic properties. Menadione is known as Vitamin K3 which is a precursor for various types of vitamin K.

The Role(s) of Natural Toxins in Predator Prey Relationships.

Except for the essentially predatory habit of a small number of carnivorous plants, predator prey relationships are normally defined as those between predatory animals and other animal species that are their prey. Natural toxins are often important components of an arsenal of chemicals used by predators to subdue prey. Where prey species are naturally more agile, and often larger than their potential consumers, toxins are employed to incapacitate them. Specific compounds produced by the predatory species may also aid in the breakdown and digestion of prey tissue. In common with the allelopathy of invasive plants and their native antagonists, the deployment of toxins by potential prey animals is often a critical defensive strategy. This is especially so for sessile organisms where stings, barbs and toxic exudates are the only means of defence against a mobile predator.

Mobile prey species have more available options which include the perception of a predator using tactile, visual and acoustic signals as well as detecting a predator's presence using olfactory sensors. Usually more than one of these senses are required to accurately assess a predatory threat. For example, olfactory indicators may alert prey to a potential threat but additional visual and tactile cues may be needed in order to trigger an effective avoidance or escape response. Predator signatures may range from the smell of a desert coyote detected by the Mohave Desert Tortoise (Nafus et al. 2017) to a waterborne olfactory signal derived from the predatory fish species Flathead Gudgeon and recognised by one of its prey, the mosquito fish (Ward & Mehner 2010). These olfactory clues therefore perform an important defensive function, as an early warning system, giving potential prey a chance to take evasive action. Whether or not these olfactory triggers include the natural toxins employed by predatory organisms to incapacitate prey, their early detection and response is certainly part of the biological warfare story.

As is the case with allelochemicals, toxic compounds can be deployed in either a defensive or attacking mode by predators and their prey. While sessile organisms can employ natural toxins as a defence against active predators, a similar strategy may be used by static predatory organisms to ambush prey. In such cases toxins produced by potential predators can even play a part in ensnaring prey through attractive pigmentation or olfactory signals.

Venoms.

Although the term venom is sometimes used synonymously with poison in the literature it is usually applied to a natural toxin that is 'delivered' by a device such as a barb, stinger or fang designed to penetrate the epithelium of a potential consumer or prey organism. Differentiation between an offensive or defensive role for a venom usually depends on the specific ecological niche of the venom producer. For example, there are over 1,200 species of venomous fish capable of inflicting painful and sometimes lethal stings to perceived predators including humans. Among the most lethal are Lionfish (*Pteroid* spp.), Stonefish (*Synanceia* spp.) and Scorpionfish (*Scorpaenidae* spp.). Most of these species rely on sharp spines coated with highly toxic, often lethal venomous exudate. Many venomous fish are relatively sedentary and adopt a defensive strategy whereby the venomous barb is concealed by a fin. Others adopt a more clearly hostile approach signified by an elaborate array of spines often accompanied by a stark pigmented display of colour; in universal language "Do Not Touch".

Defensive venoms are characteristic of a broad range of largely sessile or slow-moving invertebrate phyla such as Cnidarians (jellyfish) and Echinoderms (Sea Anemones, Starfish). However, in more active, marine molluscs such as octopuses and squid the distinction between what constitutes a venom is often blurred. A bite from the blue-ringed octopus (*Hapalochlaena* spp.) is capable of delivering a highly toxic dose of the neurotoxins dopamine and tetrodotoxin. Such a bite, although often fatal, is usually seen as defensive in nature and, as such is regarded as distinct from a venomous attack characteristic of predatory vertebrates such as snakes. Envenomation resulting from a snake bite provides a means of immobilizing an attacker or prey through the injection of neurotoxins. Such toxic compounds are capable of inducing variety of reactions including painful seizures, paralysis and respiratory failure. In addition to neurotoxins, snake venoms mainly consist of a cocktail of peptides, nucleosides, proteins, amines and electrolytes that act as cofactors in a variety of cytotoxic and haemotoxic processes. Numerous enzymes participate in this process, including phospholipases which cause the lysis of red blood cell membranes and metalloproteinases responsible for the breakdown of endothelia. For predatory snakes such tissue damage makes an important contribution to prey digestion. Proteases in snake venom may also contribute to the incapacitation of prey organisms by interfering with the blood coagulation process. Many neurotoxic peptides responsible for paralysing prey act by disrupting of the controlled passage of sodium and potassium ions through gated channels in neural membranes. This results in paralysis by inhibiting the propagation of action potentials and nervous impulses.

Although most venoms are peptides and proteins, some organisms produce a venom whose primary toxin is an alkaloid. A particularly interesting example is the American fire ant *Solenopsis invicta* which is native to Brazil and Argentina but which was probably introduced to North America via the port of Mobile Alabama and spread throughout the southern states of the U.S.A. beginning in the 1930s. Since then it has become a major pest across the southern half of the United States, capable of inflicting painful, and sometimes deadly, stings on mammals including humans. Considering that the average length of *S. Invicta* is less than 1 cm. and that they normally consume plants and insects, it might be speculated that attacks on larger animals, which are usually carried out in swarms, fall into the defensive category. However, fire ants are demonstrably capable of scavenging carcasses of dead animals so, perhaps their overall strategy may encompass elements of both offense and defence. Certainly, anyone who has come across these insects would have no problem describing them as aggressive!

As has been demonstrated many times with invasive species, and to borrow a sports analogy, "a good defence coupled with a vigorous offense is usually a good recipe for success". However, as is often the case with invasive species, the story is frequently more complex. It is now close to a hundred years since the fire ant first invaded North America from Argentina and Brazil, and its successful colonisation of the U.S.A. during that period is illustrated by its spread across the many U.S. states country from Oregon in the west to Maryland in the east. Colonisation is certainly an appropriate term for an insect noted for its elaborate colony building above and below ground, and the armies of occupants capable of aggressively expanding their territory. Individual colonies may contain as many as 200,000 individuals. Another reason for the rapid spread of *Solenopsis* in North America was their relative freedom from parasitic flies, primarily from the Dipteran order Phoridae, that acted as natural suppressors of fire ants in the South American continent but were largely absent from the fauna encountered by fire ants upon reaching North America. Examples include the parasitic Phorid flies *Apocephalus* and

Pseudoacteon that lay their eggs in the thorax of fire ants. Following migration of the fly larvae to the head of the ant this results in the beheading of the ant, an outcome that is reflected in the generic name of these parasitoid flies; derived from the Greek, A-Cephalos – without a head. The deliberate introduction of *Pseudoacteon* spp. from South America into areas of the U.S.A. infested by fire ants as a means of natural biological control has met with some success over the last thirty years, but as raised concerns over a partial switch of these parasitic flies to secondary insect prey such as honey bees that are useful as pollinators.

Unlike many ant species, which have formic acid or low molecular weight peptides as their principal active toxins, the primary active component of fire ant (*Solenopsis*) venom is the eponymous alkaloid solenopsin (Figure 31), which consists of a piperidine ring with a methyl (CH_3) and a long CH chain. Unlike many venomous toxins whose primary role is the rapid incapacitation or paralysis of prey, the poisonous nature of solenopsin relies mainly on the inhibition of a wide range of cellular and tissue functions necessary for the healthy growth and survival of prey organisms. While slower acting than many neurotoxins, fire ant venom may just as deadly. Rapid paralysis of prey would not be necessary because fire ants are capable of attacking swiftly and in very large numbers. Solenopsin has also demonstrated antibacterial activity, which is an important property in an insect noted for living in close quarters with massive numbers of its own species (colonies may contain more than 200,000 ants) together with remnants of prey organisms all within the same living space. Many of the toxic effects of solenopsin result from the inhibition of enzyme systems essential for healthy cell and tissue function. Cell differentiation, proliferation and motility as well as intercellular communication are all aspects of normal cell function modulated by a family of enzymes known as phosphoinositide 3-Kinases, which are inhibited by solenopsin. Inhibition of neuronal nitric oxide synthetases (nNOS) by solenopsin also results in a breakdown in the integrity of tissues responsible for respiratory, digestive and vascular function. Examples include the disruption of peristaltic activity and inhibition of angiogenesis which sustains and expands tissue blood vessels.

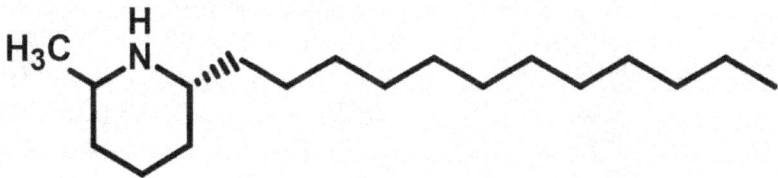

Figure 31. Solenopsin. Fire-ant venom toxin consists of a piperidine ring with a methyl (CH3) group at position 2 and a long CH chain at position 6.

Angiogenesis, the proliferation of blood capillaries supplying oxygen to tissues, is a clear indication of a healthy organism irrespective of taxonomic group. Therefore, it is not surprising that the inhibition of this process by solenopsin would ordinarily be seen as damaging to the wellbeing of biota exposed to this alkaloid. While this is true under normal circumstances, there are instances where the opposite may apply. Shrinkage and ablation of cancerous tissue may be facilitated in many cases by starving the tissue of blood supply and, therefore, oxygen. A way of achieving this is to diminish the blood supply to cancerous tissue. The solution? – inhibition of angiogenesis. The agent? – solenopsin.

The Toxicity of Trace Metals[18].

In addition to the large array of organic compounds that have long been identified as natural toxins with a huge range of biological capabilities, several metallic elements have also contributed to a rich human cultural history. In addition to the deliberate consumption of pure metals and metal-containing compounds for medical or remedial purposes, there are numerous cases of toxic metals as instruments of mischief and murder. However, most deaths and illnesses from metal intake usually result from accidental ingestion of a pesticide or solvent used for agricultural or industrial purposes.

As is the case with numerous organic chemicals many metals perform useful functions vital to the healthy metabolism and growth of biota, although, again, it is a matter of dose that separates the positive or remedial qualities of metal ingestion from potentially lethal or detrimental effects.

On the positive side, metals such as iron (Fe), manganese (Mn), magnesium (Mg), copper (Cu), zinc (Zn) and molybdenum (Mo) have been shown to act as important cofactors and coenzymes capable of enhancing the effectiveness of enzymes and other bioactive molecules. There are several ways in which enzyme activation, catalysis or enhancement may occur. For example, a metal coenzyme may play a role in facilitating enzyme access to the molecular or tissue site that is specific for that enzyme, i.e. the "active site". As such the metal might be seen as providing a key to accessing the active site, with the power of moderating a specific biological process.

In some instances, the same trace metal may perform multiple

---

[18] The term 'Trace Metal' simply refers to the fact that these elements perform biological functions in very small, i.e. trace, amounts.

roles within a particular enzyme or group of enzymes. Alcohol dehydrogenases are a family of enzymes found within a large range of organisms that are responsible for the detoxification of alcohols and their conversion to aldehydes and ketones. The trace metal zinc appears as a cofactor in this enzymic activity both as a catalyst and as an enhancer of the structural integrity of the enzyme itself.

Trace metal coenzymes and cofactors are often highly specific to the type of enzymic reaction required. For example, enzymes that employ ATP in phosphate transfer require magnesium (Mg) as an exclusive cofactor. There are many such instances of a high degree of specificity governing the role(s) of metals as micronutrients and cofactors in biological reactions. Yet, given the broad spectrum of metal atoms available to organisms and overlapping similarities in their electrochemical characteristics there also remains significant scope for interference among trace metals. This may lead to competition and displacement of critical micronutrients by other metals having similar physical and chemical properties. A potential consequence of this interference is the inhibition of the normal biochemical activity of metal cofactors leading to the dysfunction of enzymes and other biologically important compounds. Competition among metals may also have adverse effects on their abilities as electron donors and acceptors. This may have important implications for the ability of some ionic forms to function as electrolytes responsible for the passage of the electrical impulses. The physiological metals primarily involved with neuromuscular function are sodium (Na^+), potassium (K^+), calcium (Ca^{2+}) and magnesium (Mg^{2+}). Trace metal pathology can arise from competition among metals. Although trace metal toxicity may be a consequence of competition between metal coenzymes and extraneous metals, very often the toxic effects of metals on enzymes and other bioactive molecules may be unrelated to metal/metal competition. For example, ions of heavy metals such as copper, lead, mercury and silver may displace hydrogen from the cysteine residue of a proteinaceous enzyme. The resultant sulphur bond may distort the shape of the protein with the effect of inhibiting or eliminating its enzymic properties. Trace metals may be responsible for a number of sulphur-related anomalies resulting in the inhibition or malfunction of enzymes and other biologically active compounds.

How Are Metals Toxic?

In addition to the chemical features of individual trace metal compounds that define their toxicity, metals may have several different physical forms that influence their effect on biological systems.

Unlike organic compounds whose toxicity usually depends on the presence or positioning of specific chemical groups within the molecular structure, differences in metal toxicity often depend on changes to the actual atomic structure of a metal and its ionic configuration. Variations in atomic structure may involve both stable and unstable forms known as isotopes. However not all occur naturally. For example, metal radioisotopes represent a special case insofar as many are artificially synthesised to fulfil a variety of medical and industrial applications. Therefore, although many potentially toxic metal radioisotopes occur naturally, several are man-made.

In those relatively rare cases involving the exposure of biota to unstable metal isotopes, adverse effects may be a direct consequence of the radiation discharged by a specific isotope, as opposed to the chemistry of the metal itself. Isotopic instability results from a shortfall in the number of neutrons present in the metal atom, thereby affecting the atomic mass but not its chemical properties. An unstable isotope is

186

characterised by the release of nucleic energy in the form of ionic particles. The three major forms of emission associated radioactive decay are alpha, beta and gamma rays. Of these, by far the most penetrative and destructive to living systems are gamma rays. Gamma rays act by destroying the structural integrity of a many biologically active chemicals. Effects include the destruction of enzymes, structural proteins and fragmentation of nucleic acid sequences resulting in random mutations and other anomalies. By contrast, the only significant radioisotope in organic compounds is C-14, a virtually harmless weak beta emitter that represents approximately 23% of the total naturally occurring carbon in the human body, as compared to the dominant stable isotope C-12.

The toxicity and mode of action of unstable metal isotopes is outlined here primarily as a reminder that metal toxicities are often highly dependent on aspects of their physical characteristics. Even in stable isotopes of metal elements, the distribution of electrons associated with the atomic structure has a direct effect on how they interact with other charged particles or 'ions'. This ionic configuration, often referred to as 'valency' is defined by the number of unpartnered, or 'free' electrons on the outer shell of the metal atom. Bearing in mind that electrons are negatively charged, their loss will result in a positively charged ion, or cation. Thus, monovalent metal ions such as sodium (Na^+) and potassium (K^+) may possess a single positive (cationic) charge, whereas divalent metal ions e.g. calcium (Ca^{2+}) or lead (Pb^{2+}) have a double positive charge. Examples of trivalent and quadrivalent ions are aluminium (Al^{3+}) and tin (Sn^{4+}) respectively. All metals form positively charged ions.

On the other hand, anions are non-metallic, negatively charged ions, characterised by an excess of electrons. This means that, in anions, negatively charged electrons outnumber protons, thereby conferring an overall negative charge to the ion. Examples include the negatively charged ions of chlorine (Cl^-) and sulphur (S^{2-}). Bearing in mind that opposite poles (i.e. charges) attract, the valency of cations and anions defines their reactivity with each other, i.e. monovalent cations form bonds with monovalent anions; divalent cations form bonds with divalent anions.

The physiological and toxicological properties of metallic compounds rely greatly on their ionic configuration. To illustrate this point comparison may be made between two quite different compounds, lead sulphide and sodium chloride. Although lead and sodium are both metals, they have little in common. Lead itself is an element with four naturally occurring stable isotopes ranging in atomic weight between 204 and 208, with a weighted average (pardoning the pun) of 207.2. Lead is the heaviest of a group of metals given the name heavy metals for obvious reasons.

Perhaps the most similar characteristic shared by the metals sodium and lead relates to their physical softness. Sodium is a silvery white metal with a consistency not unlike hard cheese. Lead is somewhat harder, but still relatively soft and malleable compared with most metals. However, lead and sodium have little else in common. Unlike lead sulphide, which is insoluble in water, sodium chloride is very highly water-soluble; indeed, it is in its dissolved, disassociated ionic form that sodium chloride executes its physiological function(s). An integral part of this is the free cationic form (Na^+) which enables the controlled conduction electrical charge (hence the term electrolyte) that is essential for neuromuscular function.

A History of Metals and Their Uses, Both Good and Bad.

Mercury.

The pharmacological use of mercury salts was known to Paracelsus as long ago as the early sixteenth century. At that time Paracelsus was Professor of medicine at Basle in 1527 and was known to prescribe mercury salts for treatment of syphilis and intestinal disorders.

Mercury was mined at least 2,000 years ago in the Spanish Almaden mines, and the use of cinnabar (mercury sulphide, alpha-HgS) for red ink in China goes back 3000 years. It continues to have widespread industrial applications such as the manufacture of batteries, thermostats and switches as well as the universal mercury thermometer. However, its use as a biocide and a medicine has been curtailed in recent decades as a result of it toxicity.

The most toxic form of mercury is the organometallic cation methylmercury $[CH_3Hg]^+$ which is associated with the chloride or hydroxyl anion, but may also occur as dimethylmercury, $(CH_3)_2Hg$. Methylation of mercury in the natural environment usually results from the action of sulphate-reducing bacteria in a process that was first described by Jensen and Jernelov (1969). The methylation process has been described as a possible detoxification process by some authors but this remains a subject of speculation.

Methylmercury is a powerful neurotoxin responsible for a broad range of neurological and other symptoms that are particularly evident in consumer organisms. It is rapidly absorbed into the body where its high lipid solubility causes it to accumulate in nervous tissue which has a high fat content. Mercury concentration in neural cells causes a variety of cognitive disorders, cerebral palsy, ataxia and seizures. These are particularly prevalent in young children. Other recorded symptoms of methylmercury exposure have included abnormal limb growth and impaired reproductive development. The range of toxic effects on numerous tissues and organs is probably related to the ability of methylmercury to bind to sulfhydryl (thiol) groups of proteins and enzymes, thus preventing their normal function.

An important attribute that methyl mercury shares with other lipid-soluble compounds with a long chemical half-life is the phenomenon of biomagnification. This is defined as the progressive increase in concentration of the chemical i.e. mercury, as it progresses up the food chain. In other words, the top consumers in the food chain will have higher mercury concentrations than the biota at the bottom of the food chain, i.e. the primary producers. The potentially toxic effects of biomagnified chemicals such as mercury will therefore be exacerbated by a diet dominated by organisms high in the food chain.

An example of mercury biomagnification occurred in the Japanese village of Minamata where beginning in 1956, over 100 deaths were attributed to mercury poisoning through ingestion of fish contaminated with methylmercury released into Minamata Bay from an industrial source. The diet of the affected populace was dominated by fish at the apex of the food chain in the contaminated area. A feature of this incident was the high number of neural disorders and other disabilities recorded. These were particularly noticeable in newborn children resulting from the transport of methylmercury across the placental barrier in pregnant women who had consumed contaminated fish.

A major mass-poisoning incident arose in Iraq in 1971 when a consignment of grain that had been treated with methylmercury as a fungicide was diverted to make flour, which was used to make bread. Hospital records attributed 459 deaths, although it now seems probable that over 3700 deaths occurred.

Mercuric nitrate was used to treat the felt used in the hatting industry. This process (known as carroting) was used to toughen the felt fibres. This led to neurological disorders among employees in the hatting industry, vis. Lewis Carroll's 1865 novel *Alice's Adventures in Wonderland*. Note: Carroll did not invent the phrase "As Mad as a Hatter" as the term was already in common usage.

Lead.

Lead sulphide, is a naturally occurring mineral compound, generally known as galena, combining the divalent lead cation with the divalent sulphur anion $Pb^{2+}S^{2-}$. It is the chief mineral ore for lead but is also an important source of silver which co-exists with the lead ore.

In pure form lead appears as a soft malleable metal with a low melting point, qualities that have led to a variety of human applications over many millennia. However human exposure has come at a heavy price, with an estimated annual death toll from lead poisoning of more than 900,000 worldwide.

Opinions still persist that Roman neurological disorders including the bizarre behaviour of some of their notorious leaders was due to lead poisoning resulting from the use of lead for plumbing (water supply) and drinking wine from lead goblets. Supporting, but largely circumstantial evidence included the fact that lead is known to be easily malleable, and that Roman tap water at that time contained 100x more lead than local spring water. However, the notion that lead played a major role in the 'Fall of the Roman Empire' seems somewhat unlikely as collapse of Roman Empire began around 476 AD, and hard evidence of lead in Roman water supply did not emerge until 7th century. As for the more notorious, some might say imbalanced, individuals as Roman rulers, high on the list are Caligula (reigned from 37 AD – 41 AD), who was assassinated and succeeded by Claudius. Then there was Claudius' nephew Nero (reigned from 54 AD – 68 AD); Both are probably too early to implicate lead poisoning as an influence on their mentality and behaviour. Nevertheless, such stories connecting lead poisoning with the fall of Rome still persist.

Accounts of the history of lead and its use over millennia bear testament to the versatility of the metal. Despite growing evidence of the adverse effects of lead on human health it is, perhaps, surprising to learn of lead as somewhat of a delicacy for human consumption. An example is the use of lead chromate (Chrome Yellow) as a colourant in custard powder, sweets and even snuff. The addition of lead acetate to wine as a sweetener provides further, but contested, evidence of its part in the decline of the Roman Empire. However, for more tangible evidence of the dangers of lead ingestion, we cite Sir John Franklin's ill-fated 1845 expedition to find the North West Passage in North America. As their ship became stuck in ice and unable to rely on fishing and hunting other animals, deaths of all 129 crew members were attributed to eating food from cans sealed with lead solder.

Lead has a long, if somewhat dubious, history of medical application. Lead plasters were used in Roman times to treat skin conditions and as a cosmetic ingredient, and it continues to be used as a traditional medicine in some societies. In recent times lead has been used as a paint pigment, as a ceramic glaze (PbO), and in pure form as a solder. It continues to be used as an alloy and as a radiation shield in medical applications. Concerns over the inhalation of fumes of petrol containing the 'anti-knock agent' tetraethyl lead $Pb(CH_2CH_3)_4$

led to the removal of lead from petrol in the 1970s. Likewise, potential poisoning from the ingestion of flaking lead-based paint precipitated a ban on the use of lead paint in 1978. A particular concern was the widespread use of lead paint in schools where the ingestion of the metal flakes and dust by children was perceived as a potential risk to their mental development.

Cadmium.

The most serious case of cadmium poisoning from an industrial source was recorded in the Toyama Prefecture of Japan in the Jinzu River between 1910 and 1960. Water used in the mining of zinc, lead and copper was heavily contaminated with cadmium as a byproduct of the mining operation before being discharged into the river. Cadmium entered the food chain through the consumption of fish from the river and the use of river water for cooking. Water from the river was also used for the inundation of rice paddy fields. Symptoms of excess cadmium ingestion included kidney damage (nephropathy) and severe bone and joint pain. The pathology has since come to be known as Itai-Itai disease, where Itai-Itai has been roughly translated as Ouch-Ouch. Initially the outbreak was thought to be due to lead poisoning, but by 1968 the symptoms had been confirmed as the result of cadmium toxicity. Although the condition has now been largely remediated, 184 cases have been identified since 2000 with another 388 suspected cases. Kidney and bone disorders are thought to be related to the ability of cadmium to interfere with calcium metabolism through its antagonistic occupation of transmembrane calcium transporters.

Nickel.

Nickel is a silvery-white non-ferrous metal that is prized for its ability to resist oxidation, a quality that puts it in high demand as an alloy with other metals such as chromium, copper, titanium and aluminium. The corrosion- and heat- resistance of nickel alloys make them much sought after for a large range of industrial applications.

Nickel is estimated to be the fifth most abundant element on earth, although by far the largest mass of nickel is found in the earth's core. i.e. beneath the earth's crust. In the earth's crust itself the abundance nickel compares with metals such as copper and zinc. It is primarily found in Eastern Asia Indonesia, Philippines, Russia, Canada, Australia, China.

Nickel was discovered in 1754 by Axel Fredrik Cronstedt, who called it kupfernickel, or 'Devil's Copper', a name indicative of its initial confusion with copper. Nickel naturally exists as 5 stable isotopes with atomic masses between 58 and 64, and 6 radioactive forms with atomic masses ranging from 56 to 66. Radiation from Ni radioisotopes varies from soft beta to gamma radiation, although the energy output is not of sufficient strength to pose a health threat to exposed organisms.

The toxic effects of nickel are therefore related to its inherent chemical characteristics rather than its isotopic configuration. Human health hazards include allergic dermatitis and a variety of ailments following nickel ingestion, including headaches, nausea, diarrhoea and vomiting. Nickel is also classified as a carcinogen and has been linked to the development of cancerous lung tumours following inhalation.

Aside from health hazards related to human nickel ingestion and inhalation, nickel ions have been shown to play important roles as cofactors in a range of enzyme systems in plants, bacteria and fungi. The nickel-bearing metalloenzyme urease catalyses the hydrolysis of urea to ammonia and carbamate in the yeast *Cryptococcus neoformans* which is responsible for cryptococcal meningitis. This potentially fatal disease is caused by the ingestion or inhalation of spores from *C. neoformans.* There is some evidence to suggest that urease may perpetuate the cryptococcal infection and increase the virulence of the disease in infected biota. In this case we may draw the conclusion that, while this nickel enzyme favours the proliferation of the yeast *C. neoformans* within a host organism, this is not good news for the newly adopted host!

Superoxide dismutases (SOD) are a family of antioxidant enzymes whose function is to protect cells from the toxic effects of reactive oxygen species derived from superoxide. These enzymes actually fall into four categories depending on the identity of the metal cofactors. Three of these are mostly plant-based and are associated with the metals copper (Cu)/zinc (Zn) (bound to both metals), iron(Fe) and manganese (Mn). The plant sources of these SODs read like an advertisement to "eat your vegetables" – cabbage, Brussels sprouts, peas, tomatoes, spinach, chickpeas and cashews. The fourth (nickel (Ni)-containing) SOD, comprise a group found only in *Streptomyces* species and in cyanobacteria, and distinct from Mn-, Fe-, or Cu/Zn-containing SODs in terms of their amino acid sequence and metal ligand environment.

Copper.

Copper is a reddish brown nonferrous mineral which has been used for thousands of years by many cultures. The name for the metal comes from Kyprios, the Ancient Greek name for Cyprus, an island which had highly productive copper mines in the Ancient world. Copper is a good conductor of both electricity and heat, and that's why it can be found in numerous

195

electronic appliances. Copper occurs in pure metallic form in the earth's crust as well as several carbonate, silicate and sulphide compounds, where it exists primarily in the more reactive divalent (Cu^{2+}) form. Although divalent copper ions form weak inorganic complexes with carbonate and hydroxide anions, they have a tendency to form stronger bonds with organic matter where copper complexes may reach high concentrations. By contrast, the monovalent, cuprous, form (Cu^+) also occurs in the natural environment, but only very low concentrations due to its tendency to oxidise to the divalent, cupric ion in water.

Copper also binds to solid materials such as silt and oxyhydroxides of other metals such as iron and manganese. Total copper concentrations in sediment from polluted environments may exceed 5,000 mg Cu/Kg. Copper-bearing ores such as copper pyrites ($CuFeS_2$) and malachite [$CuCO_3.Cu(OH)_2$] have been known as rich sources of copper for over 6000 years. Copper mining and smelting has taken advantage of ductile and malleable properties of the metal, which in turn have led to its widespread use in plumbing and the creation of countless tools, ornaments and jewellery. The manufacture of such products has been aided by the development of harder alloys such as brass (copper-zinc) and bronze (copper-tin).

There are few recorded remedial or medical benefits specifically related to human ingestion of copper, although on the other hand copper is not regarded as a risk to the health of humans or other mammals. Indeed, some copper supplements have been recommended as additives to the food of some domestic livestock. Low copper toxicity in mammals relates in part to their ability to detoxify copper in the liver and kidney. The ability of copper to bind to organic matter is also seen as a reason for its low bioavailability from food.

Copper is, nevertheless, an essential micronutrient that is a constituent of over 30 enzymes and other bioactive compounds. These include antioxidant enzymes such as superoxide dismutases (SOD) and cytochrome C oxidase, which is responsible for the mitochondrial oxidation of nutrients releasing chemical energy via electron transport chain and through the mediation of adenosine triphosphate (ATP).

Despite its role as an important cofactor in several enzyme systems, there are several examples of copper toxicity at concentrations that exceed those required as a nutrient. For example, at high doses copper inhibits the sulfhydryl groups on the antioxidant enzyme glucose-6-phosphate dehydrogenase (G-6-PD). This reduces the capability of G-6-PD to prevent cellular damage through its free radical scavenging activity.

Other detrimental effects related to higher copper levels concern its ability to disrupt the function of synaptic vesicles in nervous systems thereby interfering with their ability to control the release of neurotransmitters. It is still debated whether copper acts by blocking or otherwise adversely affecting the function of gamma amino butyric acid (GABA) receptors or altering the structural integrity of voltage-gated Ca^{2+} channels responsible for the transmission of neural and neuromuscular impulses. Whatever the nature of the exact role of copper the relationship between copper and neural dysfunction has been linked in recent years to the neurodegenerative Wilsons disease which is characterised by the inability to detoxify dietary copper such as Menkes and Wilson diseases (Cerpa et al. 2005).

The involvement of copper in so many diverse biological processes, both positive and negative, requires a complex homeostatic mechanism that effectively transports the metal to the appropriate molecular sites responsible for its metabolic function while minimising counterproductive inhibitory activity.

An important part of copper homeostasis is an elaborate network of copper-transporting proteins and chaperone proteins also known as metallothioneins. Together, these combine to maximise the role of copper ions as micronutrients while minimising the binding of copper ions to unwanted ligands; a process that may seriously impair the function of important bioactive compounds. An example of this activity is the employment of a specific transepithelial carrier mechanism for copper wherein the substrate for the transporter is restricted to the monovalent, cuprous, form of the metal. A critical aspect of monovalent copper is its lack of reactivity, a property that allows its cellular transport in a format that minimises indiscriminate binding to bioactive compounds and ensures the safe delivery to metallothioneins.

Zinc.

Zinc is the 23rd most common element in the earth's crust, where it exists primarily as zinc sulphide, (ZnS) best known as the mineral sphalerite. It often occurs in association with ores of metals such as copper, lead, gold and silver.

The first known human use of zinc dates from about the 3rd millennium BC, when it was used in the form of alloy with copper (brass). Brass objects of various kinds have been found in several ancient civilizations in and around the Aegean Sea region, although the production of zinc on a large scale probably began with a distillation process developed in Rajasthan, India In the 9th century AD. The commercial production of zinc started in the 12th century.

Zinc was isolated and identified in pure form in 1746 by German Chemist Andreas Sigismund Marggraf, and a detailed insight in the electrochemical properties of zinc was presented by Luigi Galvani and Alessandro Volta in 1800. Zinc was named as white snow by Alchemists, as it burned in air to form white compound. The word zinc is probably derived from German word *Zinke* that means *tooth*.

Zinc is not regarded as toxic to humans in amounts usually encountered and is more widely recognised as an essential micronutrient that acts as a coenzyme and co-factor in as many as 300 enzymes. Zinc also plays a role in protein and RNA synthesis. Zinc deficiency has been diagnosed in some individuals and has been linked to symptoms as varied as diabetes, macular degeneration and prostate malfunction. However, many of these conditions have been remedied with zinc supplements.

Unlike copper, zinc does not display robust oxidative activity but nevertheless performs a wide range of biological functions in its capacity as a coenzyme and as a component of other bioactive compounds. In this multifunctional role of zinc does, however, share with copper the risk of potentially inhibitory and competitive binding to unwanted molecular sites leading to biological malfunction. It is, therefore, no surprise that, like copper, zinc is subject to an extensive regulatory process consisting of membrane proteins and transporters that control the cellular and subcellular concentrations of zinc and its movement across plasma membranes. Homeostatic control of zinc also involves its compartmentalisation into intracellular vesicles designed to ensure its availability for a myriad of important biological functions while protecting tissues from excess zinc levels.

Thallium.

Thallium was discovered by the English physicist and chemist William Crookes in 1861 who named it 'thallium' after the Greek word *thalos* meaning green shoot, which came from the characteristic green colour emitted by the metal during flame spectroscopy. Thallium was first isolated in pure form as a soft grey metal by the French chemist Claude-Auguste Lamy in 1862.

Thallium is a somewhat obscure yet highly toxic metal that is seldom regarded as a dangerous natural toxin even though its toxicity is comparable with cyanide. Its low profile as a naturally occurring toxin simply relates to the low probability of encountering the metal in toxic concentrations in the natural environment. Relatively little attention has been paid to the threat of thallium poisoning resulting from the accidental ingestion of plants and animals contaminated by thallium through the food chain. Some concern was expressed by *LaCoste et al.* (2001) over thallium accumulation by vegetables exposed to natural deposits of the metal associated with realgar (arsenic sulphide) from Macedonia. However natural mineral sources of thallium are rare and are largely confined to locations contaminated by sulphide waste derived from the mining of other metals.

The highly poisonous nature of thallium resulted in a historical market for the metal as a pesticide and fungicide. For example, during the 1930s thallium was freely available from pharmacies for the treatment of the dermal fungus Ringworm. However, in subsequent decades, an increasing incidence accidental, and even deliberate poisonings led to its decline and eventual ban for therapeutic use in several countries.

The toxicity of thallium results from its strong tendency to bind indiscriminately to the sulfhydryl groups of many proteins inhibiting their normal function. Another major source of thallium toxicity relates to its chemical similarity to potassium. This causes thallium to displace potassium at many different active sites without being able to perform the physiological roles that are characteristic of potassium itself. Essential physiological functions inhibited by thallium include the propagation and conduction of nervous impulses and the release of energy from ATP necessary for heart function. Hair growth also falls into this category. Hence the loss of hair that is one of the characteristics of thallium poisoning. Other potassium functions impaired by thallium inhibition include maintenance of ribosomal integrity, without which cell growth and maintenance are incapacitated.

Today the use of thallium is almost exclusively confined to the electronics industry where it is used in the manufacture of superconductors and photoelectric cells. It is also used as a means of altering the refractive index of optical lenses during the manufacturing process.

Despite the very low incidence of accidental poisoning due to thallium ingestion, it is impossible to ignore the significant part played by thallium in many deliberate poisonings, both fictional and real. This role is encapsulated in thallium's well-deserved title "The Poisoner's Poison"

Notable among these were a spate of Australian poisonings in the 1950s, seemingly aided by the availability of thallium in the form of rat poison, appropriately marketed under the trade name Thall-rat. Motives for these and other murders varied from monetary gain (thallium shared with the arsenic alternative title "Inheritance Powder") to revenge. However, sometimes, the term "Pathological Poisoner" seemed the 'best fit'. Such was the case of Graham Young who used thallium, antimony and other toxins to poison about 70 people between 1961 and 1971 in the English village of Bovington. The poisonings of which three were fatal apparently stemmed from a fixation with poisons exhibited by Young since childhood. Young died in Parkhurst Prison in 1990.

From the fictional standpoint it not difficult to conflate imaginary tales of 'murder by thallium' with real cases. Agatha Christie's novel, *"The Pale Horse"* written in 1961 describes as many as 10 fictional murders carried out using thallium. The author's characteristic attention to detail including victims' symptoms such as hair loss has since prompted speculation over the potential forensic value of her novel for the solution of real-life cases, i.e. 'Whodunnit?' – or perhaps inspiration for budding poisoners(?).

Antimony.

Antimony is a shiny grey-white metal is found in nature primarily as the sulphide (Sb_2S_3), known as stibnite, which is the source of its characteristic symbol, Sb.

Antimony compounds have been known for several millennia and were used as cosmetics and medicines by the ancient Egyptians as long ago as 5000 years ago. Ancient Greek physicians prescribed antimony powders for the treatment of skin disorders and the Roman scholar Pliny the Elder describes the preparation of antimony sulphide as a medication in his work *Natural History* dating from 77AD. The brittle, somewhat crystalline nature of the metal has allowed it to be ground to a powder for a variety of applications, including as both a medicine and a cosmetic.

The Arabic name kohl has been given to a mascara based on antimony sulphide. However, this product is now banned in several countries due to concern over its toxicity. As is the case with several trace metals the therapeutic and medicinal properties of antimony are often a trade-off between possible benefit and toxic risk. The possible contribution of antimony-based medication to the death of the composer Mozart is one of several case histories reflecting this delicate balance. However, accounts of the therapeutic uses of antimony compounds continued well into the 20th century. Many of these applications involved their use as treatment for parasitic worms and protozoa. For example, the malarial protozoan *Plasmodium vivax* was an early target of antimony-based remedies. Although medicinal use of antimony compounds has faded from the scene, the metalloid is still retained in some countries as a treatment for protozoan parasite *Leishmania*. Until quite recently antimony-based treatment extended to the parasitic flatworm *Schistosoma* (Trematoda), although this, too, is now curtailed because of antimony toxicity. This decline in remedial use of antimony is now widespread following a reappraisal of the health risks involved. However its non-medical use continues unabated in a variety of industrial applications.

Since the Middle Ages uses of antimony compounds and alloys have been increasingly confined to their purely metallic properties. According to some of the early metallurgy literature emerging from Italy, methods for the isolation of pure antimony metal were probably known over 600 years ago. An early description of antimony and its uses was published posthumously in 1540, in the metalworking manual *De la Pirotechnia* by the metallurgist Vannoccio Biringuccio. By then the metalloid already had several industrial applications. For example, in the mid-15th century antimony was employed a hardening agent in cast metal printing type used by Johannes Gutenberg's first printing presses. By the 1500s, antimony was used as an additive to the casting of church bells to improve their resonance.

In the 17th century a significant step in the smelting of the lead/tin alloy pewter was the addition of antimony as a hardening agent. Britannia metal as it came to be known was an alloy similar to pewter that was developed by metallurgist James Vickers in Sheffield, England in 1769. It was also called 'Vickers White Metal' after its discoverer. Unlike the original pewter composition, copper was substituted for lead in Britannia metal, a development that eliminated human exposure to lead in pewter drinking vessels.

The addition of antimony created a metal product that was more malleable than the original pewter alloy. This new-found versatility paved the way for a greatly expanded variety of

Britannia Metal products including teapots, candlesticks and vases.

In addition to ornamental uses, engineering applications for antimony continued well into the nineteenth century. A notable example was a copper, antimony and tin alloy, which was discovered in 1839 by the metallurgist Isaac Babbit. The eponymous Babbit Metal was used to create bearings that were found to reduce friction in steam engines.

Probably the largest single application of antimony was a military one. Global production of antimony reached a peak of 82,000 tons in 1916 due to its use as a constituent of projectiles in the form of bullets and shells during the First World War. The antimony/lead alloy responsible for this surge in production was originally developed back in 1784 by the British General Henry Schrapnel, whose name became synonymous with ammunition products. Following the First World War the need for antimony production again changed focus, and its primary application became that of a strengthening agent in lead-acid batteries. This function continues today. Another contemporary use for antimony is the subsidiary but synergistic effect of antimony oxide (Sb_2O_3) as a flame retardant.

When considering the many applications associated with antimony and its numerous alloys, it is not surprising that some concern has been expressed over environmental contamination related to these activities and the release of antimony in waste products (Cooper & Harrison 2009; Periferakis *et al.* 2022). Ingestion of antimony either via the gastrointestinal tract or by inhalation of mining and industrial waste have been implicated in a range of disorders many of which have symptoms similar to those associated with arsenic exposure.

Routes of antimony exposure also include the consumption of vegetables from contaminated soils that concentrate antimony through their root system. Ailments linked to antimony-based compounds include cardiovascular disorders and dermatitis.

Summary.

Stories of Natural Toxins and their cultural ramifications remain some of the enduring narratives of human history, yet their origin long pre-dates the vast majority of life on this planet. Their adoption for human use therefore represents a fleetingly small part of the evolution of these compounds, which have served to preserve and enhance the survival of the organisms that have produced them over many millions of years.

Natural Toxins have been primarily thought of as weapons of biological warfare, acting to gain a competitive advantage for their producer(s) over neighbouring biota. This has been accomplished using a variety of defensive or offensive strategies. In defensive mode Natural Toxins have been produced by biota to ward off potential consumers and predators because of their noxious odour and taste. In more extreme cases consumers are faced with sickness or death if they are exposed to Natural Toxins produced by potential prey or food organisms. The defensive capability of these compounds often extends to antimicrobial activity designed to combat infestation and parasitism of the host organism. This property has made Natural Toxins and their derivatives particularly attractive as antibiotics for medical and agricultural use. However, antibiotic properties are just one of many reasons why these compounds have been sought after and developed for human use.

Natural Toxins display an enormous range of bioactive properties whose functional characteristics are often dictated by the presence of specific chemical groups as part their molecular structure. These bioactive compounds are commonly biproducts of metabolism that are not generally part of the primary metabolic processes such as locomotion and digestion. Instead, these secondary metabolites undergo transformation by specific enzymic pathways into compounds that enhance the survivability of the parent organism. This is generally achieved through interaction with other biota and as a response to changes in the surrounding physical environment. Despite not possessing enzymic properties themselves, natural toxins may act as co-factors for enzymic function. Part of this activity relates to their ability to strengthen and sustain the integrity of proteins as well as other structurally important molecules such as polysaccharides.

Natural Toxins also perform a variety of physiological functions many of which have an impact on the nervous systems of organisms that are exposed to these chemical compounds. For example, nitrogen-containing cyclic compounds known as alkaloids comprise numerous neurotoxic and psychoactive compounds. Many of these interfere with the passage of electrolytes across nerve membranes and neural junctions such as synapses and specific receptor sites in the central and peripheral nervous system. These effects result from their ability to mimic the normal neurotransmitters responsible for the controlled passage of nerve impulses. The type of effect resulting from the occupation of a specific receptor site by a foreign chemical depends on whether the compound acts as a protagonist or antagonist of the normal substrate. In other words 'will the occupation of a receptor site by a copycat chemical enhance or block the normal function of the neural receptor site or junction?'.

Although not generally regarded as having a biological origin, inorganic chemicals including trace metals may also display characteristics of natural toxins because of their roles as enzyme co-factors and as inhibitors of a broad range of physiological processes. Inhibitory effects on biochemical pathways often result from the substitution of a functional group or element by a metal ion or metallic compound that is incapable of carrying out the normal biological activities that characterize a healthy, unexposed organism.

Bacterial methylation of some metals, notably mercury, probably satisfies the definition of a natural toxin as a substance of biological origin since the methylation transformation is instigated by living organisms, probably as part of a detoxification process that protects the microorganisms responsible for the methylation process.

However, included here are several metals whose toxic effects largely result from man-made industrial products and the human effort of extracting metals from the natural environment, Some metalloids, such as arsenic have a rich human history that incorporates both industrial and more sinister practices. As such they illustrate both the positive and negative characteristics of so many natural toxins that find their way into the environment. Usually the difference between beneficial (e.g. medicinal) properties and toxic (e.g.

poisonous) effects of these natural compounds is simply a matter of the amount ingested; the dose. Hence, the principal theme of this book, "Natural Toxins: The Good, the Bad and the Deadly".

References.

Bennett, J.W. and M. Klich (2003). *Mycotoxins.* Clinical Microbiol. Rev. 16(3), 497–516. Doi:10.1128/CMR.16.3.

Campbell, P.G.C., P.V. Hodson, P.M. Welbourn and D.A. Wright (2022). *Ecotoxicology.* Cambridge University Press. 576pp. ISBN: 9781108819732.

Cerpa, W., Varela-Nalar, L., Reyes, A.E., Minniti, A.N. and N.C. Inestrosa (2005). Is there a role for copper in neurodegenerative diseases? Mol. Aspects Med. 26, pp. 405-420. Doi:10.1016/j.

Cooper, R.G. and A.P. Harrison. (2009). *The exposure to and health Effects of antimony.* Indian J. Occup. Environ. Med. 13, 3-10. Doi. 10.4103/0019-5278.50716.

Dawkins, R. (1989*). The Selfish Gene.* Oxford University Press. (2nd Edn.), Oxford, U.K.

Eagleson, M. *Concise Encyclopedia Chemistry.* 1st Edn. Berlin: Walter de Gruyter; 1994.

Eiswerth, M.E., T. D. Darden, W.S. Johnson, J. Agapoff, and T.R. Harris. (2005). *Input-output modeling, outdoor recreation, and the economic impacts of weeds.* Weed Science, 53:130-137.

Enders, M., and J.M. Jeschke, (2018). A network of invasion hypotheses. In: J.M. Jeschke & T. Heger (Eds.), *Invasion biology: Hypotheses and evidence.* (pp. 49-59). Wallingford, U.K.: CABI.

Harissis, H.V. (2014). *A Bittersweet Story: The True Nature of the Laurel of the Oracle of Delphi.* Persp. Biol. Med. 57 (3) pp. 351-360.

Hartmann, T. (1999). *Chemical ecology of pyrrolizidine alkaloids.* Planta 207, pp. 483–495.

Holland, A. & S. Kinnear. (2013*). Interpreting the possible ecological role(s) of cyanotoxins: compounds for competitive advantage and/or physiological aide?* Mar. Drugs. 11, 2239-2258.

Jensen, S. and A. Jernelov (1969). *Biological methylation of mercury in aquatic organisms.* Nature, 223, 753-754.

Kaur, R. and S. Arora. (2015). *Alkaloids. Important therapeutic secondary metabolites of plant origin,* J Crit Rev. 2(3): 1–8.

La Coste, C., B. Robinson and R. Brooks (2001). *Uptake of thallium by vegetables: Its significance for human health, phytoremediation and phytomining.* J. Plant Nutr, 24, 1205-1215.

Murray, S.A., T.K. Mihali and B.A. Neilan. (2011). *Extraordinary conservation, gene loss, and positive selection in the evolution of an ancient neurotoxin.* Mol. Biol. Evol. 28, 1173-1182.

Nafus, M.G., J.M. Germano, and R.R. Swaisgood. (2017). *Cues from a common predator cause survival-linked behavioral adjustments in Mojave Desert tortoises (Gopherus agassizii).* Behav. Ecol. Sociobiol. 71, 1-10.

Nowacki, E. & D. Nowacka. (1965). *Biogenetic Classification of Alkaloids.* (English translation). In: Wiadomosci Botaniczne. (Wiad. Bot.) 9(3), 207-216.

Panaccione, D.G. (2023). *Derivation of the multiply-branched ergot alkaloid pathway of fungi.* Microbial Biotechnology. 16, 742–756. doi: 10.1111/1751-7915.14214

Pelletier, S.W. (1983) *The nature and definition of an alkaloid. Alkaloids:* Chemical and Biological Perspectives. Wiley.

Periferakis A., Caruntu A., Periferakis A.-T., Scheau A.-E., Badarau I.A., Caruntu C. and C. Scheau. (2022). *Availability, Toxicology and Medical Significance of Antimony.* Int. J. Environ. Res. Public Health. 19:4669. 29pp., doi: 10.3390/ijerph19084669.

Pimentel, D., S. McNair, J. Janecka, J. Wightman, C. Simmonds, C. O'Connell, E. Wong, L. Russel, J. Zern, T. Aquino and T. Tsomondo. (2001). *Economic and environmental threats of alien plant, animal and microbe invasion.* Agriculture, Ecosystems & Environment. 84(1), pp. 1-20.

Setzler-Hamilton, E.M., D.A. Wright and J.A. Magee (1997). *Growth and spawning of laboratory-reared zebra mussels in lower mesohaline salinities. Chapter 9. In; Zebra mussels and aquatic nuisance species.* Ed. F.M. D'Itri, Publ. Ann Arbor Press Inc., Chelsea, Michigan, U.S.A.

Ward, A.J.W. and T. Mehner (2010). *Multimodal mixed messages: the use of multiple cues allows greater accuracy in social recognition and predator detection decisions in the mosquitofish, Gambusia holbrooki.* Behavioral Ecology, 21(6), 1315–1320.

Wright, D.A. and J.A. Magee (1997). *The toxicity of endod extract to the early life stages of Dreissena bugensis.* Biofouling, 11(4), pp. 255-263.

www.ingramcontent.com/pod-product-compliance
Lightning Source LLC
Chambersburg PA
CBHW070302290326
41930CB00040B/1805